Global Environmental Studies

For further volumes:
http://www.springer.com/series/10124

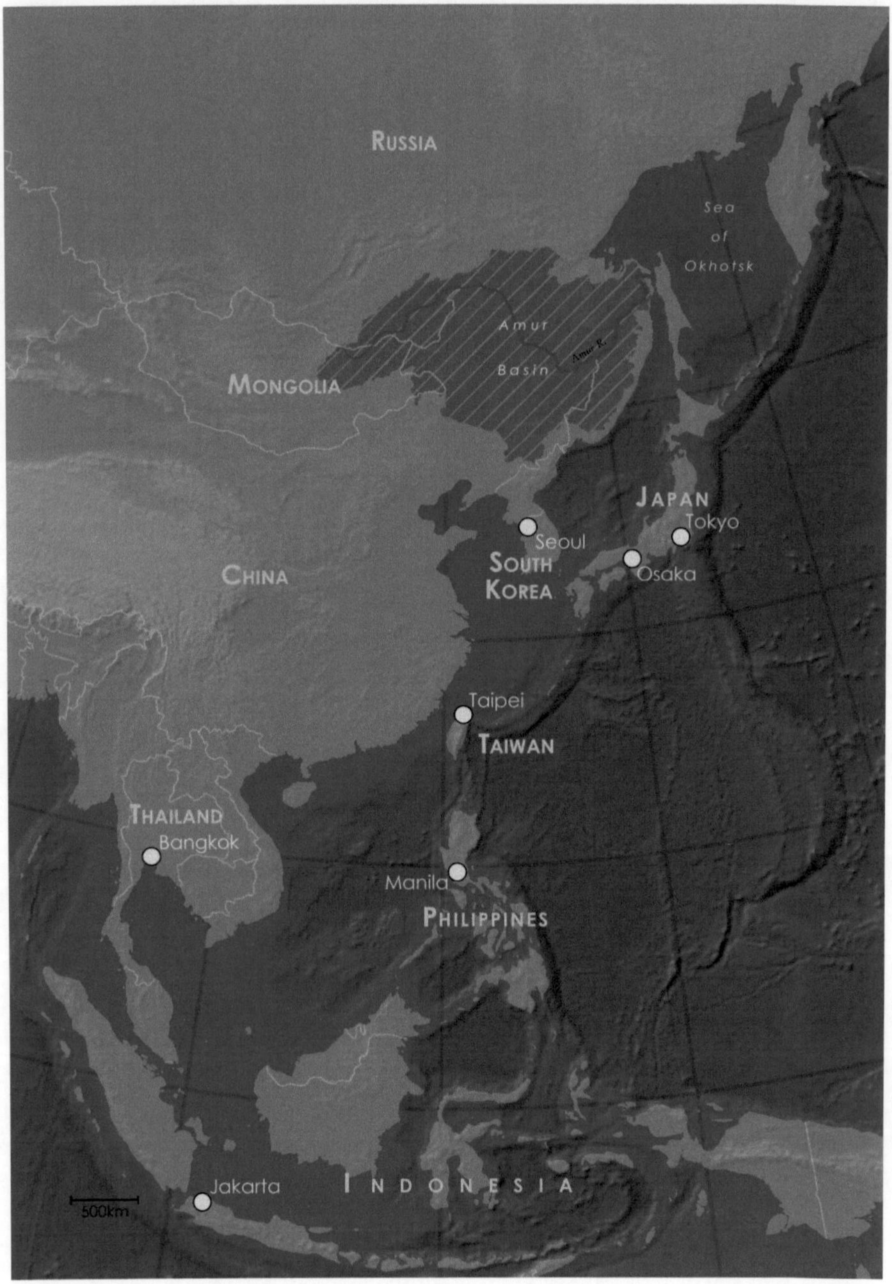

Russia
Sea of Okhotsk
Amur Basin
Amur R.
Mongolia
China
Japan
Tokyo
Seoul
South Korea
Osaka
Taipei
Taiwan
Thailand
Bangkok
Manila
Philippines
Jakarta
Indonesia
500km

Makoto Taniguchi • Takayuki Shiraiwa
Editors

The Dilemma of Boundaries

Toward a New Concept of Catchment

Springer

Editors
Makoto Taniguchi, D.Sc.
Research Institute for Humanity and Nature
457-4 Motoyama, Kamigamo
Kita-ku, Kyoto 603-8047, Japan

Takayuki Shiraiwa, D.Sc.
Institute of Low Temperature Science
Hokkaido University
Kita-19, Nishi-8
Kita-ku, Sapporo 060-0819, Japan

ISSN 2192-6336 ISSN 2192-6344 (electronic)
ISBN 978-4-431-54034-2 ISBN 978-4-431-54035-9 (eBook)
DOI 10.1007/978-4-431-54035-9
Springer Tokyo Heidelberg New York Dordrecht London

Library of Congress Control Number: 2012935863

Printed on acid-free paper

Springer is part of Springer Science+Business Media (www.springer.com)

Preface

Boundaries have broad meanings: There are natural boundaries of two different spheres, conditions, or materials such as land-and-ocean and surface-and-subsurface environments; social boundaries such as national and local demographic boundaries; and human boundaries such as those created by races, cultures, and religions. These boundaries are necessary for maintaining each society and for managing each environment. For instance, there are individual laws for each area, sphere, or condition, and each individual country has different laws on the usage of water on land. In addition, different laws have been applied to water under the ground (groundwater), even in the same country. The management of water in the ocean also differs from the management of water on land. Although water circulates seamlessly above the surface as vapor, on the ground as surface water, under the ground as groundwater, and in the ocean as sea water, different ways and laws exist to control each type of water separately.

There are many water problems depending on boundaries, which include land subsidence and trans-boundary of water movement beyond the demographic boundaries. Land subsidence is caused mainly by excessive groundwater pumping, and this relationship demonstrates the situation of the "tragedy of commons." Individual people use cheap (free) groundwater instead of relatively expensive tap water to get individual benefits. However, the accumulation of individual small benefit by consumption of groundwater can lead to land subsidence, which will then cause all individual benefits to be lost. The solution for land subsidence by regulation of pumping may be the symptomatic solution. It was used in Tokyo and Osaka in the 1970s and might help solve similar problems that latecomer cities, such as Bangkok, Jakarta, and Manila, are facing. However, what should be understood fundamentally is that land subsidence occurs because of the tradeoff between water fluxes as resources and water stock in subsurface environment. We need boundaries to manage the water; however, those boundaries cause many problems because water moves seamlessly. This is the "dilemma of boundaries."

Water movements across boundaries cause not only problems but also benefits. Transports of dissolved materials from land to the ocean feed nutrients to the life in the sea. This concept is so called "fish-breeding forest" through ecological services (linkages) between river basin and ocean. This continental-scale terrestrial–marine linkage, the giant fish-breeding forest, is another example of the dilemma of the boundaries, because it is a trans-boundary issue across countries with different stakeholders and also an interdisciplinary study between hydrology and oceanography beyond the boundary between land and ocean.

Water on land and ocean is a "commons," and it moves seamlessly beyond boundaries. However, we still do not know how to manage the water beyond the dilemma of the boundaries and the tragedy of commons. In this book, we will discuss how the boundaries have important roles and meanings and cause many problems not only in water and material cycles, but also in laws and institutions. Finally, we summarize the dilemma of the boundaries toward a new concept of the basin.

Kyoto, Japan Makoto Taniguchi
Sapporo, Japan Takayuki Shiraiwa

Contents

Contributors

Hermanni Backer Helsinki Commission, Helsinki, Finland

William C. Burnett Department of Earth, Ocean and Atmospheric Sciences, Florida State University, Tallahassee, FL, USA

Natasha Dimova Department of Earth, Ocean and Atmospheric Sciences, Florida State University, Tallahassee, FL, USA

Takahiro Endo Graduate School of Life and Environmental Sciences, University of Tsukuba, Tsukuba, Ibaraki, Japan

Yasunori Hanamatsu Slavic Research Center, Hokkaido University, Sapporo, Japan

Takahiro Hosono Priority Organization for Innovation and Excellence, Kumamoto University, Kumamoto, Japan

Osamu Ieda The Slavic Research Center, Hokkaido University, Sapporo, Hokkaido, Japan

William Todd Jarvis Department of Geosciences, Institute for Water and Watersheds, Oregon State University, Corvallis, OR, USA

Juha-Markku Leppänen Finnish Environment Institute, Marine Research Centre, Helsinki, Finland

Osamu Matsuda Graduate School of Biosphere Sciences, Hiroshima University, Higashi-Hiroshima, Japan

Stephen McCauley George P. Marsh Institute, Clark University, Worcester, MA, USA

Humio Mitsudera Institute of Low Temperature Science, Hokkaido University, Hokkaido, Japan

Takeshi Murota Department of Economics, Doshisha University, Kyoto, Japan

Takanori Nakano Research Institute for Humanity and Nature, Kyoto, Japan

Takeshi Nakatsuka Graduate School of Environmental Studies, Nagoya University, Nagoya, Aichi, Japan

Takeo Onishi River Basin Research Center, Gifu University, Gifu, Japan

Shammy Puri Secretary General, International Association of Hydrogeologists, Dorchester on Thames, Oxon, UK

Jun Shimada Graduate School of Science and Technology, Kumamoto University, Kumamoto, Japan

Takayuki Shiraiwa Institute of Low Temperature Science, Hokkaido University, Sapporo, Japan

Makoto Taniguchi Research Institute for Humanity and Nature, Kyoto, Japan

Keisuke Uchimoto Institute of Low Temperature Science, Hokkaido University, Hokkaido, Japan

Hiroshi Wakana Muroran Institute of Technology, Muroran, Hokkaido, Japan

Chusei Yamada Special Assistant to the Minister for Foreign Affairs, Government of Japan, Tokyo, Japan

Global Environmental Series

The Global Environmental Studies series introduces the research undertaken at, or in association with, the Research Institute for Humanity and Nature (RIHN). Located in Kyoto, Japan, RIHN is a national institute conducting fi xed-term, multidisciplinary, international research projects on pressing areas of environmental concern.

RIHN seeks to transcend the common divisions between the humanities and the social and natural sciences, and to develop synthetic and transformative descriptions of humanity in the midst of a dynamic, changeable nature. The works published in the series will refl ect the fullbreadth of RIHN scholarship in this transdisciplinary field of global environmental studies.

Editorial Structure

Part I
Introduction

Chapter 1
Introduction

Makoto Taniguchi and Takayuki Shiraiwa

Abstract This chapter shows a brief introduction of four sections which are summarized as ignored linkages between surface and subsurface environments, trans-boundary linkages of land and ocean, the impacts of human-made boundaries, and challenges for new management beyond boundaries. This chapter brings readers to understand the importance of ignored linkages and how to redraw a traditional concept of catchment management into a new one.

Keywords Boundaries • Surface-subsurface interaction • Land-ocean interaction • Trans-boundaries • Ignored linkage

Although water circulates continuously and seamlessly on Earth, various research areas such as oceanography, surface hydrology, groundwater hydrology, climatology and glaciology are usually undertaken separately. However, recent findings related to interactions of water in land, oceans, and the atmosphere encourage researchers to more comprehensively understand the behavior of water through collaborative works that go beyond the boundaries of each discipline (e.g. Taniguchi et al. 2008).

Water is also separated by numerous human-made boundaries such as national borders (Fig. 1.1), vertical administrative systems and so on. When these human created boundaries disrupt natural water circulation, water-related environmental problems can be invoked, and/or it gets very difficult to cope with these issues.

M. Taniguchi (✉)
Research Institute for Humanity and Nature,
457-4 Motoyama, Kamigamo, Kita-ku, Kyoto 603-8047, Japan
e-mail: makoto@chikyu.ac.jp

T. Shiraiwa
Institute of Low Temperature Science, Hokkaido University,
Kita-19, Nishi-8, Kita-ku, Sapporo 060-0819, Japan

M. Taniguchi and T. Shiraiwa (eds.), *The Dilemma of Boundaries: Toward a New Concept of Catchment*, Global Environmental Studies,
DOI 10.1007/978-4-431-54035-9_1, © Springer Japan 2012

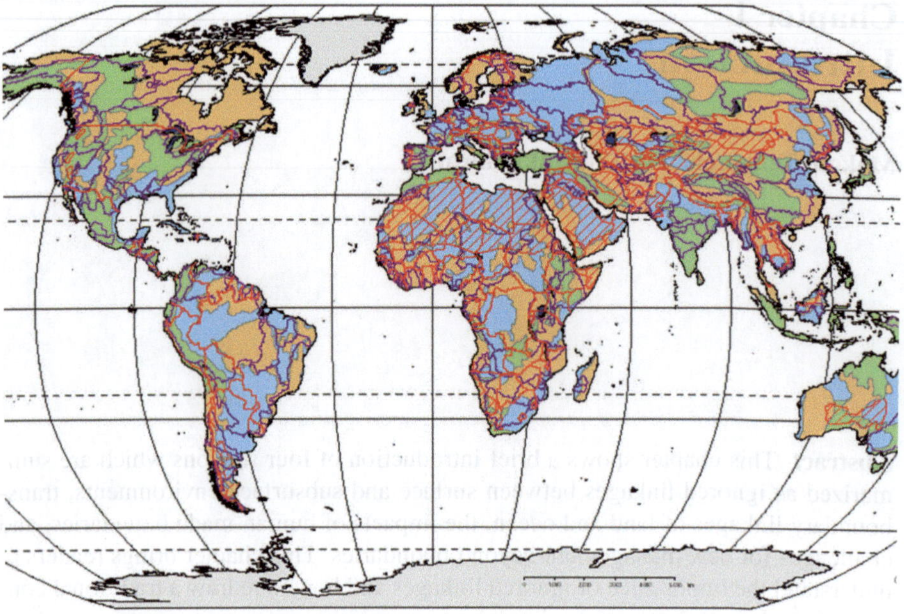

Fig. 1.1 Transboundary of groundwater (ISRAM, 2006)

Thus, artificial boundaries in natural water circulation raise important topics that need to be reconsidered, particularly in the social sciences.

In this book, we especially focus on two boundaries; one is that between surface water and groundwater; the other is that between basin water and ocean water. The book consists of four sections on these two boundaries: (1) ignored linkages between surface and subsurface environments, (2) trans-boundary linkages of land and ocean, (3) the impacts of human-made boundaries, and (4) challenges for new management beyond boundaries.

In sections 1 and 2, we consider the importance of interactions between surface water and groundwater, and between land water and ocean water, mainly based on findings from the natural sciences. The RIHN project "Human impacts on the urban subsurface environment (Urban Subsurface Environment (USE) project, PL: Makoto Taniguchi)" shows the importance of interactions between the surface environment and the subsurface environment in terms of water, material and thermal transports. The RIHN project "Human activities in Northeastern Asia and their impact on biological productivity in the North Pacific Ocean (Amur-Okhotsk (AOP) project, PL: Takayuki Shiraiwa)" introduces the new idea of a "giant fish-breeding forest," which shows the importance of interactions between land and ocean.

In the USE project (Chap. 4), seven Asian cities are investigated to evaluate how deep and fast human impacts have affected the subsurface environment during last 100 years, crossing the boundary between surface and subsurface environments

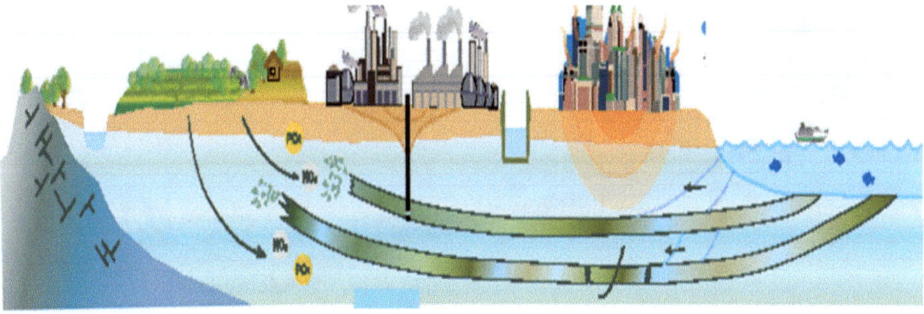

Fig. 1.2 Urban subsurface environment project

(Fig. 1.2). Cumulative human impacts on subsurface environments were documented at depths of up to 200–300 m. Groundwater circulation was accelerated by more than ten times in the past century due to groundwater pumping, again showing between surface and subsurface environments. Subsurface thermal storage due to surface warming, such as by the urban 'heat island effect,' is 2–6 times that attributable to global warming. Numerical modeling of the subsurface environment in Tokyo, Osaka, Bangkok, and Jakarta allowed evaluation of groundwater recharge rates and area, residence times, and exchanges of fresh/salt water between land and ocean. Creation of a 0.5 km grid GIS database based on nine categories of land cover/use in three different historical periods (1930s, 1970s, and 2000s) allowed evaluation of water, materials, and heat exchange between surface and subsurface environments in each city (Taniguchi 2010, 2011).

Takahiro Hosono (Chap. 5) introduces the nitrate-arsenic (NA) boundary as an important concept in aquatic environmental studies. This NA boundary is proposed in the USE project, which revealed the importance of different geological and geomorphological constructs as an important new boundary concept for catchments. It is hoped that this concept will be used to help construct a water management system that will reach beyond the borders of particular countries in Asia in which arsenic pollution is a major problem.

William Burnett (Chap. 6) presents a model to assess groundwater inflows to surface water via tracer techniques. Radon (Rn) can serve as a relatively rapid and inexpensive groundwater tracer. The techniques are applied in Asian coastal areas to evaluate water and material exchanges between land and ocean, and the approach works well for small areas.

In the Amur-Okhotsk Project (AOP) described above, the new global environmental concept of the "giant fish-breeding forest" (GFBF) is created by expanding the traditional Japanese idea of Uotsuki-Rin (fish-breeding forest), which related the upstream forest with the coastal ecosystem both physically and conceptually (Chap. 8). The AOP found that primary production in the Sea of Okhotsk and in the Oyashio region depended on dissolved iron transported from the Amur River and its

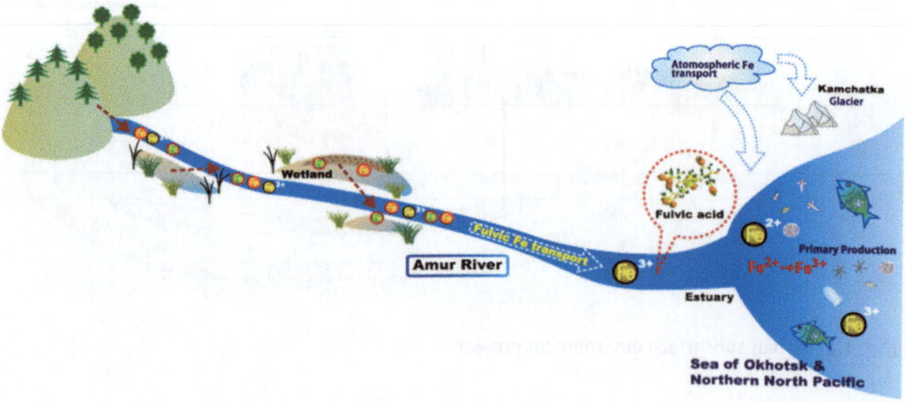

Fig. 1.3 Amur-Okhotsk project

watershed (Fig. 1.3). Therefore, the Amur River basin can be recognized as the GFBF of the Sea of Okhotsk and the Oyashio region. This hypothesis presents new perspectives on global environmental issues: an ecological linkage between the continent and the open sea, relationships between less dependent stakeholders in the system, and the identification of environmental common ground across coast lines and complex international boundaries (Shiraiwa 2010, 2011).

Takeo Onishi and others (Chap. 9) assessed the transport of dissolved iron in the GFBF system based on numerical modeling of terrestrial and marine processes. They succeeded in simulating dissolved iron production and transport, leading to the claim that further wetland conversion to agricultural lands would have significant impact on dissolved iron flux although it was to early to conclude the direct impact on the primary production in the Sea of Okhotsk and the Oyashio region.

Next, in sections 3 and 4, we consider the impacts of human-made boundaries, mainly based on findings from the social sciences. Human-made boundaries, such as national borders or vertical administrative systems are necessary constructions for the human organization and resource management issues, however the differences between the boundaries and natural boundaries such as basins, is a cause of many environmental problems. For instance, not only international rivers but also international groundwater flows, are trans-boundary issues which require a new framework for management. Lower level socio-political units, such as prefecture or city boundaries, also present similar trans-boundary problems. The section discusses both international and national/local level boundary issues.

With the outcomes of the AOP in mind, Yasunori Hanamatsu emphasizes the importance of coordinating existing legal systems and policies in an integrated manner in such a heterogeneous region crossing national as well as regime boundaries (Chap. 12).

The problems of land–ocean linkages may be usefully addressed by recalling a traditional environmental concept introduced by Hiroshi Wakana (Chap. 13). He explains how the Japanese developed the idea of "Uotsukirin" (fish-breeding forest). According to Wakana, the concept can be dated back to tenth century Tokushima Prefecture, in western Japan. The idea of "Uotsukirin" then spread throughout Japan by seventeenth century.

In Sect. 14.3, Todd Javis (Chap. 14) shows the importance of integrating groundwater boundary concerns into catchment management. He concludes that (1) groundwater boundaries can extend beyond catchment boundaries, (2) groundwater boundaries differ from catchment boundaries based on values, (3) differing values for groundwater create overlapping boundaries and conflict, and (4) managing common pool resources like groundwater requires "blurring" the boundaries.

Takahiro Endo (Chap. 15) highlights a boundary between surface water and groundwater in the Japanese legal system and discusses the role of this legal boundary in promoting the efficient use of water. In Japan, while surface water is subject to public regulations, groundwater belongs to land ownership. This boundary causes inefficient use of water resources when externalities occur between surface and groundwater users.

In Sect. 19.4, Shammy Puri (Chap. 19) declares the trans-boundary nature of aquifer issues in the title *Aquifers know no boundaries—but farmers do!* Boundaries are sometimes seen but many times unseen, and therefore the scientific community seems to communicate across boundaries, both of seen and unseen. Cross community collaboration requires patience and perceptive. A globalised world disregards boundaries—the flow of capital transcends boundaries, for example—yet a globalised world insists on boundaries for self preservation.

Jun Shimada (Chap. 20) shows an example of the trans-boundary management of groundwater resources in the Kumamoto area, Japan. Reuse of abandoned previous rice paddies for an artificial groundwater recharge area beyond the city border is revealed as a trans-boundary groundwater management issue. The success of groundwater management in this case was underpinned by robust knowledge of the hydro-geological setting and the education of local citizens about their groundwater.

Based on the successful history of the famous Helsinki Commission's ecosystem-based approach for the Baltic Sea, Juha-Markku Leppanen demonstrates how regional environmental cooperation can evolve over the years and how quantitative targets can effectively support trans-boundary management for the marine environment.

Osamu Matsuda introduces the Japanese system of management of in the coastal marine environment (Chap. 18). The establishment of the Basic Ocean Law in 2007 and the Basic Ocean Plan in 2008 introduced comprehensive management of watershed and coastal waters as a concept. It faces difficulties, however, due mainly to bureaucratic sectionalism. While citizen and NGO/NPO activities aimed at connecting the functions of forests, rivers and the sea are becoming more popular, the Satoyama and Sato-umi connections attempted in the Seto Inland Sea are promoting a new type of integrated coastal and watershed management.

Finally, in the Chap. 21 which concludes the book we try to redraw the traditional 'catchment' concept into a new one that might overcome the disadvantages caused by present artificial boundaries.

References

ISRAM (2006) Groundwater regions, political boundaries, river basins & non recharged aquifers, http://www.transboundarywaters.orst.edu/publications/register/images/asia.gif

Shiraiwa T (2010) Final report of the Amur Okhotsk project 2005–2009. In: Shiraiwa T (ed) Report on Amur-Okhotsk project No.6, 3-19. http://www.chikyu.ac.jp/AMORE/2010 proceedings6/02.shiraiwaT_3-19.pdf

Shiraiwa T (2011) Fish-breeding forest as global environmental studies. Syowado, 231p (in Japanese)

Taniguchi M (2010) Subsurface environment in Asia, -forbitten global environmental problem-. Gakuho-sha, 246pp

Taniguchi M (2011) Groundwater and subsurface environments, -Human impacts in Asian coastal cities-. Springer, 312pp

Taniguchi M, Burnett WC, Fukushima Y, Haigh M, Umezawa Y (2008) From headwater to the ocean—hydrological changes and management. Taylor & Francis, London, 650pp

Chapter 2
Codification of International Law for Transboundary Aquifers by the United Nations

Chusei Yamada

Abstract One of the important functions of the United Nations General Assembly is to codify international law in order to establish the rule of law required for the justice and order of the international community. The UN International Law Commission is the subsidiary organ of the UN General Assembly which prepares the basic documents for such codification. The Commission embarked on the work of formulating draft articles on the law of transboundary aquifers in 2002. The author was appointed as the Special Rapporteur for this project and the Commission was able to report its final result to the UN General Assembly in 2008. The UN General Assembly is to make its decision on how to transform these draft articles into the treaty. Most States share transboundary aquifers with their neighbours. Groundwater is an indispensable life support resource for mankind and also the most heavily exploited single resource for which no alternative exists. The establishment of legal norms for the proper management of transboundary aquifers, protection of the environment, and international cooperation and the peaceful settlement of disputes among States sharing aquifers is urgently required by the international community. This forward is designed to present the background of the issue, explain how the draft articles are formulated in cooperation with hydrogeologists and groundwater administrators, and describe the main elements of the draft articles, the positions of various States and the future prospect of the draft articles.

Keyword International Law for Transboundary Aquifers

C. Yamada (✉)
Special Assistant to the Minister for Foreign Affairs, Government of Japan,
3-16-1 Ebisu, Shibuya-ku, Tokyo 150-0013, Japan
e-mail: chuyama@mtd.biglobe.ne.jp

M. Taniguchi and T. Shiraiwa (eds.), *The Dilemma of Boundaries: Toward a New Concept of Catchment*, Global Environmental Studies,
DOI 10.1007/978-4-431-54035-9_2, © Springer Japan 2012

In order to secure justice and order and to settle any dispute among States through peaceful means, it is essential to establish the "rule of law" in the world community. The sources of such law in international law are treaties and international customary law. Treaties bind their States' parties only. On the other hand, customary international law binds all the States of the international community. Customary international law is defined as an "international custom, as evidence of a general practice accepted as law" in Article 38 of the Statute of the International Court of Justice. Historically speaking, in the absence of a world legislature, international law has largely developed as customary law. However, it is sometimes difficult to define what the customary rules are and there also often exist differences of interpretation of such rules among States. Furthermore there exist many lacunae in customs. In order to remove such ambiguities, there have been efforts to restate existing customary rules as agreements between States. This process is referred to as the codification of international law. The Charter of the United Nations, in its Article 13.1(a), provides that it is one of the important functions of the UN General Assembly to promote the codification of international law.

The UN International Law Commission (hereafter referred to as ILC) was established in 1947 as a subsidiary organ of the UN General Assembly with a mandate to prepare the basic documents in the form of draft articles for such codification. The ILC currently consists of 34 members who have recognized competence in international law and are elected by the UN General Assembly, bearing in mind that in the ILC as a whole representation of the main forms of civilization and of the principal legal systems of the world should be assured. The United Nations has so far adopted many important codification treaties on the basis of the works of the ILC in such fields as Diplomatic and Consular Relations, Law of Treaties, Law of the Sea and Jurisdictional Immunity.

With respect to freshwaters, the Rhine and the Danube Rivers were subject to international regulations as early as in the beginning of the nineteenth century to ensure free navigation on those rivers. In fact, the river commissions which were established for regulating navigation on these international rivers were the precursors of the present international administrative organizations. The first time the United Nations dealt with transboundary freshwater resources was when it instructed the ILC in 1970 to take up the study of the law of non-navigational uses of international watercourses. Since the mid-twentieth century, large projects have been undertaken along the various international rivers of the world, including the construction of dams and other activities for the purposes of drinking, power generation, and irrigation, among others. These activities have threatened to cause adverse effects upon downstream States. To regulate these activities, the United Nations adopted in 1997 the Convention on the Law of the Non-Navigational Uses of International Watercourses on the basis of the work of the ILC. While that convention theoretically covers such groundwaters as are physically linked to international surface waters, it was essentially meant to regulate surface waters. I will come back to this point later. In preparation for that convention, the ILC did discuss the question of whether to include groundwaters in the project. Though it recognized the need to deal explicitly with groundwaters, it decided that a separate study is required for that purpose. Meanwhile, the United Nations became aware of the rapidly expanding exploitation of groundwaters for portal, industrial and irrigation uses in both developed and developing countries and of the resulting critical overexploitation and pollution problems.

Accordingly, the United Nations instructed the ILC in 2001 to proceed with the work on "Shared Natural Resources" which were generally understood to include groundwaters, oil and natural gas. The ILC embarked on the work in 2002, appointing me as its Special Rapporteur for this new topic. Though there exist many similarities between groundwaters on one hand and oil and natural gas on the other, there are also many dissimilarities between them. Upon the recommendation of the Special Rapporteur, the ILC chose to adopt a step-by-step approach by embarking first on the work on groundwaters as the follow up to the 1997 convention on international watercourses. However, the codification work on the law of transboundary aquifers required a multi-disciplinary process. As stated before, the ILC is a body composed solely of lawyers of public international law. It does not possess scientific and technical knowledge of groundwaters or expertise for the proper management of these aquifers. The United Nations Educational, Scientific and Cultural Organization (UNESCO) is the coordinating agency of the UN organizations on world water issues. It mobilized a team of groundwater scientists, groundwater administrators and water lawyers to assist the ILC. The team included Mr. Shammy Puri, Secretary General of the International Association of Hydrologists who has also authored a chapter in this book. Their untiring and valuable support was critical in helping the ILC formulate the draft articles. During the work with the experts, it was found that international lawyers and experts often employ different concepts and terminologies. Great efforts have been exerted in selecting the language to be used in the draft articles so that they could be commonly understood by scientists, administrators, lawyers and by ordinary peoples.

The draft articles formulated by the ILC are based on the scientific evidence provided by the experts. The ILC identified ample State practices and almost 400 relevant treaties—general, regional and bilateral—on the basis of which customary rules could be identified. The States have also shown keen interest in the ILC work as aquifers exist in almost all States and the overwhelming majority of States possess transboundary aquifers with their neighbouring States. Many States transmitted valuable inputs and observations to the ILC. Taking into account the advice of experts and observations from governments, the ILC formulated a final set of 19 draft articles on the law of transboundary aquifers in 2008. The text of the draft articles is appended to this statement. The ILC has also made available the detailed commentaries to each of the draft articles. They are contained in Chapter IV of the ILC report of 2008 (UN 2008) which can be downloaded from the UN web-cite.

The ILC took 24 years to complete the formulation of the draft articles on the Non-Navigational Uses of International Watercourses. It was rather a rare case for the ILC that the codification work on transboundary aquifers was completed in such a short period of 6 years. It shows that the ILC was fully aware of the current critical situation of groundwaters and of the urgent need to establish a legal framework for the proper management of transboundary aquifers in order to achieve the objectives of equitable and reasonable utilization, protection of environment and international cooperation.

The UN General Assembly received the draft articles favourably. I am convinced that the General Assembly recognized that the draft articles are not only scientifically and technically sound but also incorporate the positions of the majority of the member States of the United Nations. It adopted the resolution 63/124 entitled "the Law of transboundary aquifers" by consensus on December 11, 2008. The resolution encouraged the States concerned to make appropriate bilateral or regional arrangements for the proper management of their transboundary aquifers, taking into account the provisions of the draft articles. It further decided to include in the provisional agenda of its 66th session in 2011 an item entitled "the Law of transboundary aquifers" with a view to examining, inter alia, the question of the form that might be given to the draft articles.

The salient points of the draft articles are as follows:

2.1 Article 1

The scope of the application of the draft articles is (a) the utilization of transboundary aquifers, (b) other activities that have or are likely to have an impact upon such aquifers, and (c) measures for the protection, preservation and management of such aquifers. The concept of other impacting activities, in the paragraph (b), was not included in the case of the 1997 watercourses convention. In the case of aquifers, activities other than utilization above them, such as those causing pollution to the aquifers or those which harm the normal functioning of aquifers by blocking or destroying the geological formation of the aquifers, must be regulated. I will come back to this point later when I take up Article 6.

2.2 Article 2

Article 2 is the definition article. Aquifer means both a geological formation which serves as a container and the water contained in the saturated zone of the formation. It is necessary to include the geological formation in the definition of aquifer in order to preserve proper functioning of aquifers. It is also necessary to include the geological formation in order to regulate aspects of its utilization such as storage, disposal of waste or new experimental techniques for carbon dioxide sequestration.

2.3 Article 3

I have received critical comments from some international lawyers for the inclusion of the sovereignty clause in Article 3, which might in their view diminish the value of the whole exercise. I do share some of their apprehension. However, we must squarely face the current state of affairs. It was the UN which passed the resolution "Permanent sovereignty over natural resources" 1803(XVII) in 1962. Many aquifer States insisted on the inclusion of the sovereignty article. It is noted that the second sentence of the Article states that it [the aquifer State] shall exercise its sovereignty in accordance with international law and the present draft articles. I believe that the current draft Article 3 represents an appropriately balanced text.

2.4 Article 4

One of the essential principles is the equitable and reasonable utilization of aquifers. Factors relevant to such equitable and reasonable utilization are listed in Article 5. The principle of equitable utilization among States sharing the same resources is identical to the term invoked in the case of the watercourses convention. However the principle of reasonable utilization is quite different here. The principle of sustainable utilization can apply only to renewable resources. International law has developed the precise legal concept of sustainability in relation to marine living resources. You find the principle of sustainable utilization in almost all fisheries conventions. This principle is clearly defined as "to take measures, on the best scientific evidence, to maintain or restore populations of harvested species at levels which can produce the maximum sustainable yield" in Article 119.1(a) of the UN Convention on the Law of the Sea (UNCLOS). It requires users to maintain the size of the population of a particular stock of fish such that the fishery can produce the maximum catch year after year. Science tells us that such a level is somewhat below the maximum population of a particular fish stock which the nature can support. This principle could be applied to other renewable resources. The 1997 international watercourses convention applied this principle and defined it as "optimal and sustainable" utilization in its Article 5.1. It meant that the watercourse States are obliged to limit the amount of use of water to that of recharge and keep the river flowing permanently. When we learnt the dynamics of aquifers, it became clear that this sustainability principle could not apply to both recharging and non-recharging aquifers. For non-recharging aquifers, there is no room whatsoever to apply this principle, as any utilization would lead to depletion of the resources. Even for recharging aquifers, recharge is, in most cases, just a fraction of the large volume of waters accumulated over hundreds and thousands of years and States could not be deprived of the use of such accumulated resources while States with non-recharging aquifers were free to use them. Accordingly Article 4 does not refer to sustainability at all and provides that a recharging aquifer shall not be utilized at a level that would prevent continuance of its effective functioning. Meanwhile, however, the term "sustainability" has become a sort of catch phrase for many environmentalists. Taking into account their positions, the term "sustainable development" is inserted in Article 7, the General Obligation to Cooperate.

2.5 Article 6

Another important principle is the obligation not to cause significant harm to other States. This is the cardinal principle of international law. This principle applies not only to adverse effects to other States caused by utilization of aquifers but also to adverse effects on the aquifers of other States caused by other activities as defined in subparagraph (b) of Article 1. Kyoto, Japan, the location of the conference which inspired this book, has a huge aquifer underneath the city. This aquifer supported the city as the old capital for 1000 years. In the 1960s, the people came to notice the lowering of the water table in some parts of the city, which coincided with the construction of subway networks. There might be cases that such infrastructure construction can be classified as adverse other activities. However, utilization of aquifers and other activities are activities necessary for society. Other States therefore have the obligation to bear certain harm as long as such harm does not go beyond the level of significant harm. The concept of "significant" is relative and cannot be defined in abstract. However, in view of the fragility of aquifers and the difficulty of removing pollutants from aquifers once they are affected, the threshold of "significant" is much lower than in the case of surface waters. In relation to this article, Article 15 also provides for a consultation procedure for any planned activity which may affect a transboundary aquifer and thereby may have a significant adverse effect upon another State.

2.6 Article 7

Yet another important principle is that of international cooperation. The key to the proper management of aquifers is international cooperation among aquifer States. The draft articles provide various measures beginning with the regular exchange of data and information, monitoring, and the establishment of joint management mechanisms.

2.7 Article 16

The draft articles also regulate non-aquifer States. In particular, Article 16 requires all States to promote technical cooperation with developing States in the scientific, educational, technical, legal and other fields for the protection and management of aquifers. Countries such as Japan, although a non-transboundary aquifer State, would be able to play a significant role in this field.

The relationship between the present draft articles and the 1997 international watercourses convention and other treaties are left for future negotiation when the final status of the present draft articles become definite. It is generally understood that the 1997 international watercourses convention covers groundwaters which are linked to international watercourses. Its Article 2(a) provides that "Watercourses mean a system of surface waters and groundwaters constituting by virtue of their physical relationship a unitary whole and normally flowing into a common terminus." It is not clear what kind of relationship between surface waters and groundwaters is envisaged here. The present draft articles cover all transboundary groundwaters regardless of whether they are recharged by or discharged to international watercourses. It must also be pointed out that according to the definition of the present draft articles, recharge and discharge zones are outside of the aquifers. The ILC considered that all groundwaters possess distinct characteristics different from those of surface waters. For instance, the Nubian Sandstone Aquifer System is linked to the River Nile south of Khartoum. However the bulk of the Nubian system has only the characteristics of a non-recharging aquifer. Accordingly, that aquifer system must be regulated by the present draft articles. The regulations of the present draft articles are generally much broader and stricter than the 1997 international watercourses convention. Dual application of the two instruments to a particular aquifer would normally not cause any difficulty. However, if it does, we would have to draft a provision to regulate the relationship between the two instruments.

There are 2 years to go before the UN General Assembly makes the final decision on the status of the draft articles. The best outcome would be to transform the draft articles to a UN convention as done in the case of the 1997 UN international watercourses convention. If it turns out to be difficult, the second best outcome would be for the UN General Assembly to adopt the draft articles as guidelines. There are certain differences in legal effects between the two alternatives. However, that difference would not matter much. As long as the draft articles receive an official endorsement by the UN General Assembly, the States concerned could make full use of the draft articles in negotiating bilateral or regional agreements with their neighbouring States in order to properly manage their transboundary aquifers.

Our task is twofold. First, the grave situation of many aquifers must be highlighted. Second, understanding and appreciation of the draft articles must be promoted. UNESCO's International Hydrological Programme (UNESCO-IHP) will hold a series of regional and general meetings in this regard. The valuable support of scientists and experts to this endeavour will be critical.

References

United Nation (2008) Report of the International Law Commission, Sixtieth session (5 May–6 June and 7 July–8 August 2008), A/63/10, 2008, 402p

Yamada C (2010) International management of transboundary aquifers—recommendation from the United Nations. JAHS 40(3):71–84

Part II
Ignored Linkage Between Surface and Subsurface Environments

Chapter 3
Introduction

Takanori Nakano

Abstract Geochemical dilemma in coastal aquifer system: lesson from history.

Keywords Groundwater • Eutrophication • Arsenic • Neutrient • Radioisotope • Stable isotope

3.1 Groundwater: The Key to Understanding the Linkage

The linkage between surface and subsurface environments and between the land and the oceans is now well understood. Water circulates in each of these environments, but far less is known about the existence and flow of subsurface water than is known about surface water and seawater. Thus, the key to understanding these linkages is improved knowledge of the role of groundwater. Chapter 4 demonstrates that the rapid expansion of human activities in Asian mega-cities has affected the quantity and quality of available groundwater. Chapters 5 and 6 each present examples of the application of geochemical tools to solve groundwater problems.

3.2 Estimation of Water Balances Using Geochemical Methods

The sustainable use of groundwater requires sound knowledge of recharge and discharge rates of aquifers; however, estimation of these rates is generally difficult. Chapter 4 discusses the groundwater pumping history of several Asian mega-cities.

T. Nakano (✉)
Research Institute for Humanity and Nature,
457-4 Motoyama, Kamigamo, Kita-ku, Kyoto 603-8047, Japan
e-mail: nakanot@chikyu.ac.jp

M. Taniguchi and T. Shiraiwa (eds.), *The Dilemma of Boundaries: Toward a New Concept of Catchment*, Global Environmental Studies,
DOI 10.1007/978-4-431-54035-9_3, © Springer Japan 2012

Over pumping has caused subsidence in many areas. However, regulation and prohibition of excessive pumping has generated other problems, such as the floating of subway constructions when pumping has been reduced. When the underground environment has been changed by human activities, it is difficult to return it to its original state. This is an important lesson that we should learn from the history of the use of groundwater resources.

Chapter 4 also shows that nutrient loads, such as silica and phosphorous, delivered to the ocean by groundwater are almost equivalent to those delivered by river water. It is important to understand this when we evaluate the contribution of human activities to the eutrophication of marine ecosystems. The flux of nutrients in groundwater to the ocean is a function of nutrient concentration and discharge rate. Thus, precise estimates of the water balance in aquifers, particularly recharge and discharge rates, are essential not only for wise use of groundwater, but also for evaluation of the effects of groundwater discharge to the oceans.

Groundwater discharge is generally determined by seepage meters. However, a seepage meter can determine an accurate discharge rate for only for the small area around its location. The spatial variation of discharge rates is considerable. Chapter 6 demonstrates a way to evaluate groundwater discharge by using radon isotopes. He estimated the flow rate of groundwater into Lake Haines, a small lake in Florida, USA for which there is no input of river water. He successfully modeled groundwater input to the lake by comparing the concentration of radon in the atmosphere, which changed according to the weather, with that in lake water, which was relatively homogeneous and independent of the weather. When there is river water discharge to lakes or the ocean, several additional factors must be taken into consideration to apply his model. However, tracer approaches such as this method are useful for determining discharge rates of groundwater.

3.3 The Water Quality Dilemma

Radon is formed by the decay of radioisotope of uranium 238 (^{238}U). The concentration of uranium ions is affected by oxic and anoxic conditions of water. The concentration is high in oxic environments, whereas it is low in anoxic ones. Similarly, N in the form of nitrate ions (NO_3^-) is soluble in oxic water, providing high NO_3^- concentrations, but NO_3^- concentrations are low in anoxic water because of denitrification processes. Excessive concentrations of NO_3^- can cause problems such as eutrophication in aquatic systems and methemoglobinemia in human infants. Large amounts of NO_3^- discharged into the aquatic environment by human activities (e.g., sewerage and chemical fertilizer) cause widespread nitrate pollution in big cities all over the world, and in Asian countries in particular.

Arsenic pollution also is a serious problem in many countries (Nordstrom 2002). Chapter 5 discusses As and NO_3^- pollution in Asian cities. Concentrations of As in water are high in anoxic conditions, whereas those of NO_3^- are high under oxic

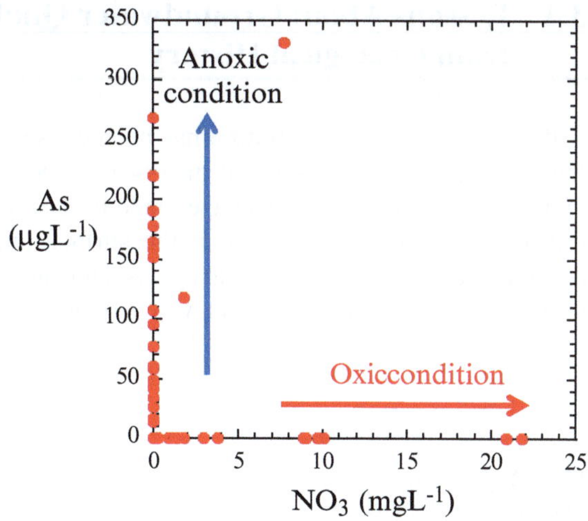

Fig. 3.1 Relationship between NO_3^- and As in the shallow groundwater of Bangladesh (calculated from data of Nickson et al. 2000)

conditions. Chapter 5 proposes an index for the environmental quality of groundwater based on the combinations of both As and NO_3^-.

Arsenic pollution is at serious levels in Bangladesh and is becoming a matter of concern on many coastal plains in southeastern Asia. Figure 3.1 shows the relationship between NO_3^- and As concentrations in the shallow groundwater of Bangladesh. As shown by Chap. 5, NO_3^- concentrations in Bangladesh are high under oxic conditions, whereas As concentrations are high under anoxic conditions. Thus, pollution problems exist in both oxic and anoxic groundwaters.

This is the dilemma for the management of water quality.

To overcome this dilemma, scientific studies on the geochemical behavior of the components of groundwater systems are a prerequisite. Many detailed studies on As pollution in Bangladesh have reached the following conclusions. Arsenic in nature is generally associated with pyrite (FeS_2), when As is substituted for S in the pyrite molecular structure (e.g., FeAsS, arsenopyrite). Pyrite is the most common naturally occurring sulfide. In Bangladesh, some pyrite occurs in roughly spherical clusters of microcrystals known as framboidal pyrite. Framboidal pyrite can form in anoxic marine environments. Iron (Fe) is not soluble in oxic water, but dissolves in anoxic water to form ferrous ions (Fe^{2+}). Seawater contains high concentrations of sulfate ions (SO_4^{2-}), which, under anaerobic conditions are reduced into bisulfide ion (HS^-) by the activity of sulfate-reducing bacteria. This bisulfide ion easily reacts with dissolved Fe^{2+} to form pyrite, sometimes as framboidal pyrite, which scavenges As. Thus, As is firmly substituted for sulfur in the mineral structure; therefore, as long as pyrite is stable under anoxic condition, the As it contains will not be released into groundwater.

3.4 Lessons About Groundwater Quality from Geological History

Although groundwater pollution has been caused by human activities over the last few thousand years, geological history provides invaluable information about the more recent changes humans have made to the subsurface environment. Figure 3.2 illustrates the natural changes to the surface and subsurface environments of a coastal area in response to glacial cycles. During the last Pleistocene interglacial age (1 in Fig. 3.2), coastal sediments were submerged and framboidal pyrite containing

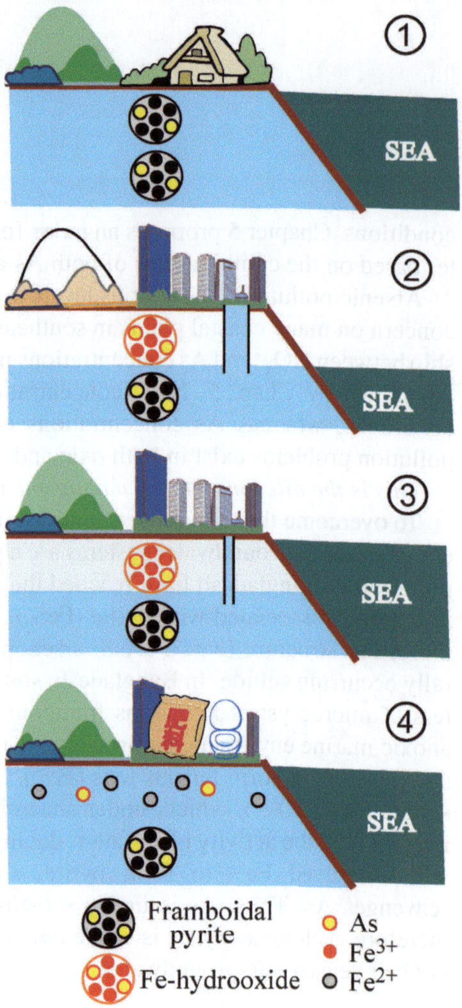

Fig. 3.2 Schematic illustration on the analogy between the behavior of As and Fe in coastal groundwater in response to sea-level changes since the last interglacial age and the effect of human activities (pumping and nutrient load) on the subsurface environment

As was formed in an anoxic environment. During the following glacial age (2 in Fig. 3.2), marine regression lowered sea level to about 100 m below present sea level. The marine regression created an aerobic environment in shallow sedimentary basins in coastal areas, and pyrite was oxidized to form hydroxides. During the subsequent (present) interglacial age, sea level rose again, submerging the previously formed Fe-hydroxides and incorporating them into the groundwater (3 in Fig. 3.2). Arsenic is strongly adsorbed onto the surface of Fe-hydroxides, but is insoluble in an oxic environment. Accordingly, as long as oxic conditions prevail, As is not released into the groundwater. However, if groundwater is returned to a reducing environment, where Fe-hydroxides are unstable, the absorbed As is released into the groundwater (4 in Fig. 3.2).

Under anaerobic conditions, Fe-hydroxides become unstable and Fe^{3+} dissolves in groundwater to form Fe^{2+}. Importantly, in addition to the increased concentrations of Fe, there are increases in Mn and As. Arsenic is toxic, but as shown by Dr. Shiraiwa's Amur-Okhotsk project, Fe is an essential nutrient for the support of primary marine production. This parallel behavior of arsenic and iron provides another dilemma for the management of water quality.

The above general model explains the behavior of As and Fe in groundwater in response to changes in subsurface environments. However, Chap. 20 questions the contribution of human activities to the changes of As concentration in groundwater. The geological history described above sheds some light on this question. From the geochemical point of view, groundwater pumping allows the introduction of air to the subsurface environment, generating oxic groundwater, which is analogous to the glacial age (2 in Fig. 3.2). Likewise, the recovery of depleted groundwater levels corresponds to the subsurface environment of the interglacial age (3 in Fig. 3.2). Transfer of nutrient loads from the surface to the subsurface environment as a result of human activities causes groundwater eutrophication, which in turn creates favorable conditions for the formation of reducing environments. Anaerobic groundwater resulting from eutrophication accelerates the desorption of As from Fe-hydroxides (4 in Fig. 3.2). Thus, the subsurface environment will not return to its original state as long as humans carry out excessive pumping of groundwater and introduce nutrients to it. The degradation of groundwater quality is of similar importance to the decline in quantity shown by pumping histories.

3.5 To Overcome the Water Quality Dilemma

In addition to the nutrient loads introduced by humans, global warming and local heating due to urbanization increase biological activity, bacterial activity in particular. These human influences may also affect the anaerobic subsurface environment, contributing to the emergence of water quality problems.

The subsurface environment is composed of sediments and hard rocks. The materials dissolved in groundwater are strongly affected by biological activity as well as by minerals such as iron sulfides and iron oxides. It is thus important to study the linkages among water, minerals, and biological activity. This requires the integration of studies in hydrology, geology, and biology.

How can we link biota, water, and minerals? Individuals have their own distinct elemental compositions. Stable isotopes can provide element fingerprints, but isotope studies in the environmental field have in the past been limited by the need for expensive equipments and specialized analytical skills. However, during the past two decades automated instruments that can simultaneously determine stable isotope ratios for many elements and samples have been developed. We can now apply stable isotope methods in environmental studies and use them for environmental monitoring. At RIHN, state-of-the-art mass spectrometers have been installed. The multiple stable isotope analysis methods now available can contribute to solving the dilemmas of water quality management and their use as a conventional analytical tool will lead to cooperative research across several disciplines.

References

Nickson RT, McArthur JM, Ravenscroft P, Burgess WG, Ahmed KM (2000) Mechanism of arsenic release to groundwater, Bangladesh and West Bengal. Appl Geochem 15:403–415

Nordstrom DK (2002) Worldwide occurrences of arsenic in groundwater. Science 396: 2135–2136

Chapter 4
Linkages Beyond the Surface–Subsurface and Land–Ocean Boundaries for Better Environmental Management in Asia

Makoto Taniguchi

Abstract Coastal cities in Asia were established and developed with little thought of their significance to surrounding subsurface and water environments. The Research Institute for Humanity and Nature (RIHN) project "Human impacts on subsurface environment" has shown important surface–subsurface and land–ocean linkages in the Tokyo, Osaka, Bangkok, Jakarta, Manila, Seoul, and Taipei basins. The relationships between city development stage and various subsurface environments in Asia were evaluated beyond the boundary between surface and subsurface environments. To partition earth surface water, thermal energy and materials into above- and below-surface components, land use/cover changes at three periods (the 1930s, 1970s and 2000s) in seven Asian cities were analyzed using GIS with a 0.5 km grid resolution. Urbanization reduced water transfer between the surface and subsurface environments and the land and ocean environments through decreasing groundwater recharge into aquifers and groundwater discharge into the ocean. Saltwater intrusion due to excessive pumping and thermal transport into the subsurface due to heat island effect increased after urbanization. Documenting the linkages of water, heat and materials between the surface–subsurface and land–ocean environments improves understanding of the full dynamism of the biogeophysical and geochemical flows that sustain the biosphere, and can improve urban planning, particularly in relation to critical groundwater and coastal resources.

Keywords Groundwater-surface water interaction • Heat island • Land cover/use changes • Land–ocean interaction • Saltwater intrusion • Submarine groundwater discharge • Urbanization

---

M. Taniguchi (✉)
Research Institute for Humanity and Nature,
457-4 Motoyama, Kamigamo, Kita-ku, Kyoto 603-8047, Japan
e-mail: makoto@chikyu.ac.jp

M. Taniguchi and T. Shiraiwa (eds.), *The Dilemma of Boundaries: Toward a New Concept of Catchment*, Global Environmental Studies,
DOI 10.1007/978-4-431-54035-9_4,

4.1 Introduction

Populations continue to increase in Asian coastal cities, which already contain a large percentage of the world's total population. Such cities cause increased water demand and material/heat discharge, and they are located at both the surface–subsurface and land–ocean boundaries (Fig. 4.1). In this situation, the impacts of such cities on their surrounding environments and the likely affects of these local impacts on other environments, are of key concern.

Increased demand on urban groundwater resources has caused subsurface environmental problems such as land subsidence due to excessive pumping. After groundwater pumping in many countries was regulated, water use switched from groundwater to surface water. Although subsurface water is connected with surface water via the hydrological cycle, the two systems have been treated separately in both the natural and social sciences.

Subsurface environmental problems were long ignored because of the invisibility of the problems and the difficulty of subsurface evaluations. Subsurface environmental problems, such as land subsidence due to excessive groundwater pumping, subsurface thermal anomalies, and groundwater contamination, have occurred repeatedly in major Asian cities, depending on the development stage of urbanization (Taniguchi et al. 2009a; Lieberman 2003; Ness and Prem 2004).

Recent global warming is widely considered a global environmental issue only above the ground. However, subsurface temperatures are also increasing because increased surface temperature penetrates the surface. In addition to global warming, the heat island effect, in which there are higher air temperatures in city centers than in surrounding area, creates subsurface thermal contamination in many cities. The combined heat island and global warming effects reach down to more than 100 m below the land surface, with the rate of increase due to the heat island effect being much higher than that caused by global warming (Taniguchi et al. 2007b).

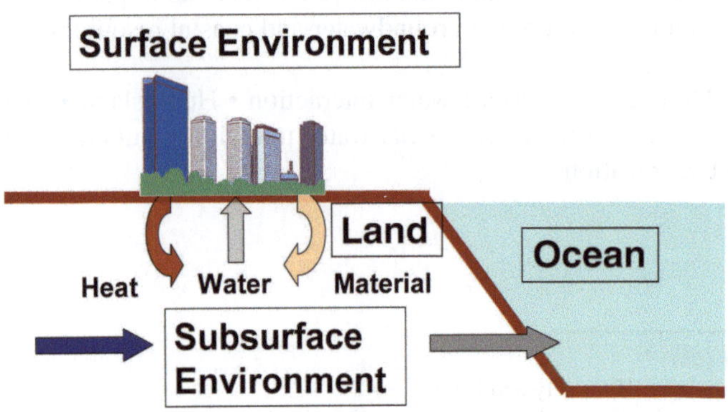

Fig. 4.1 Land–ocean and surface–subsurface linkages

Another conventional boundary for water and material transport exists between the land and ocean. Saltwater intrusion is a subsurface environmental problem between the land and ocean boundary. Excessive groundwater pumping in coastal cities decreases subsurface groundwater potential, which induces salt water to penetrate from the ocean. Regarding material (contaminant) transport to the coast, direct groundwater discharge has been recognized recently as a significant pathway for both water and material from land to ocean (Taniguchi et al. 2002; Burnett et al. 2003). Many major Asian cities are located in the coastal zone, which makes material and contaminant transport by groundwater a key to understanding coastal water pollution and the effects on associated ecosystems.

4.2 Study Areas and Methods

In the Research Institute for Humanity and Nature (RIHN) project "Human impacts on subsurface environments", the dynamics of surface–subsurface and land–ocean boundaries were analyzed for four domains (urban areas, groundwater, subsurface heat, and subsurface contamination) (Fig. 4.1). Intensive field observations and data collections were made in the Tokyo, Osaka, Bangkok, Jakarta, Manila, Seoul and Taipei basins (Taniguchi et al. 2009a, 2009b; Fig. 4.2).

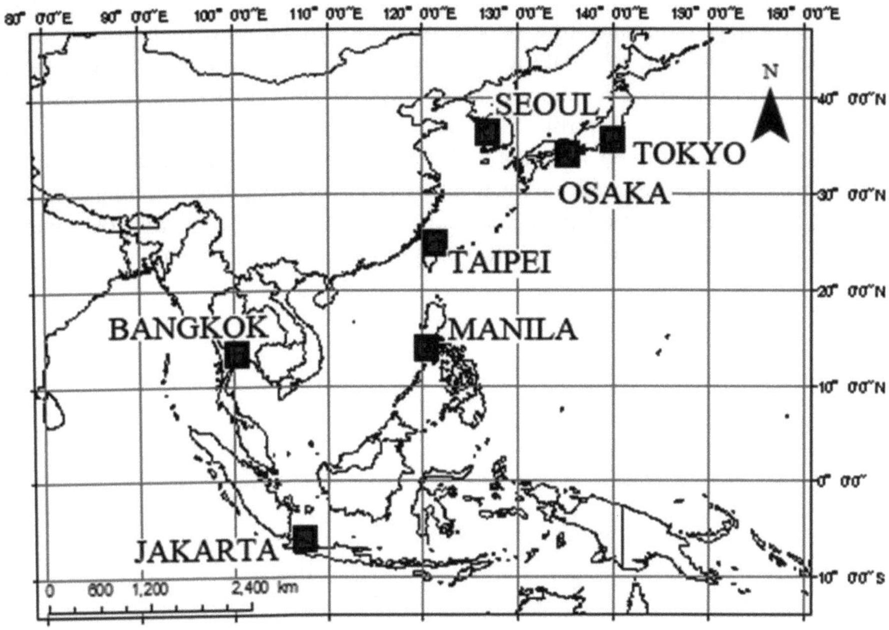

Fig. 4.2 Location of the study area

Surface environmental properties which were analyzed included air temperature, surface water quantity/quality, and degree of urbanization. Subsurface environments were observed intensively for groundwater, subsurface heat, and subsurface contamination. Subsurface temperatures were measured in boreholes at 1-m depth intervals to evaluate climate change and heat island effects (Taniguchi and Uemura 2005, 2007a; Yamano et al. 2009). Groundwater stable isotopes (δO and δN) and chemical compositions were analyzed to evaluate the groundwater flow system and the origin of the contamination (Umezawa et al. 2009). Numerical simulations based on the US Geological Survey's modular flow model (MODFLOW) were performed in Tokyo, Osaka, Bangkok, and Jakarta to evaluate water exchange between land and ocean. Material transport by groundwater was also evaluated using seepage meters and Radon (Rn), which can measure direct groundwater discharge into the ocean (Burnett et al. 2007).

To evaluate the land surface environment responsible for distributing water, heat and materials at the earth surface into the subsurface, land-use/cover changes for the seven targeted cities in three periods (1930s, 1970s and 2000s) were analyzed using a Geographic Information System (GIS) with 0.5-km grid resolution.

4.3 Linkage Between Surface Water and Groundwater

Changes in reliable water resources between groundwater and surface water reservoirs occurred in many Asian cities, depending on urbanization development stage. In the early stage of urbanization, people used more groundwater than surface water because of safety and affordability. Severe land subsidence due to groundwater pumping then occurred in many coastal cities, because their aquifers consisted mainly of sedimentary deposits (Fig. 4.3). For example, land subsidence in the Osaka plain was observed since the 1930s due to excessive groundwater pumping for industrial water uses, causing the local government to regulate pumping after the 1960s. In Bangkok, land subsidence was observed in the 1970s, and the government started regulating pumping in the late 1990s (Taniguchi et al. 2009b).

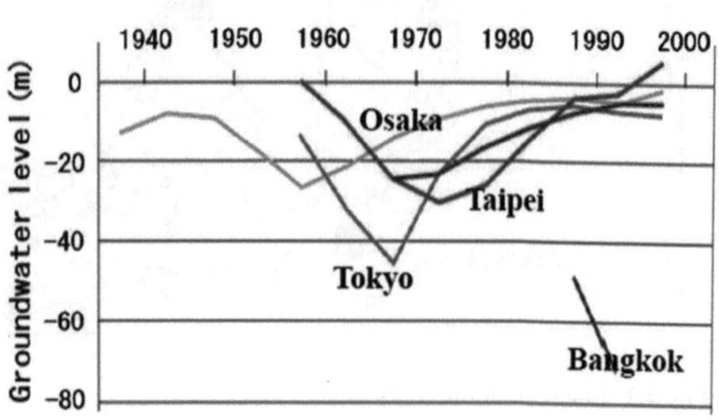

Fig. 4.3 Changes in groundwater level above sea level in Asian cities

Groundwater is "private water" which belongs to the land owner, while surface water is "public water", which is managed publicly. Therefore, from the legal and bureaucratic perspective, groundwater is managed differently than surface water, even though both waters are connected hydrologically. This causes many water management problems beyond the surface and subsurface boundary (Taniguchi et al. 2009d).

4.4 Linkage Between Surface and Subsurface Thermal Environments

The effect of surface warming on the subsurface thermal environment was also evident. Subsurface temperature is affected by both global warming (Huang et al. 2000) and the heat island effect due to urbanization (Taniguchi and Uemura 2005). The combined effects of these two processes reach down to more than 100 m below the land surface, and can have potential consequences on groundwater systems (Taniguchi et al. 2007b). Subsurface temperatures in four Asian cities (Tokyo, Osaka, Bangkok and Seoul) were compared in order to evaluate the effects of surface warming due to urbanization and global warming, with regard to the developmental stage of each city (Taniguchi et al. 2007b). Mean surface warming in each city ranged from 1.8°C to 2.8°C. The depth of deviation from the regional geothermal gradient was deepest in Tokyo (140 m), followed by Osaka (80 m), Seoul (50 m), and Bangkok (50 m). Analysis of the timing of the start of surface warming showed that the depth of 0.1°C deviation from a constant geothermal gradient in subsurface temperature was deeper when the elapsed time from the start of surface warming due to urbanization was longer. This trend was confirmed by air temperature records from the study areas during the last 100 years (Taniguchi et al. 2007b; Fig. 4.4).

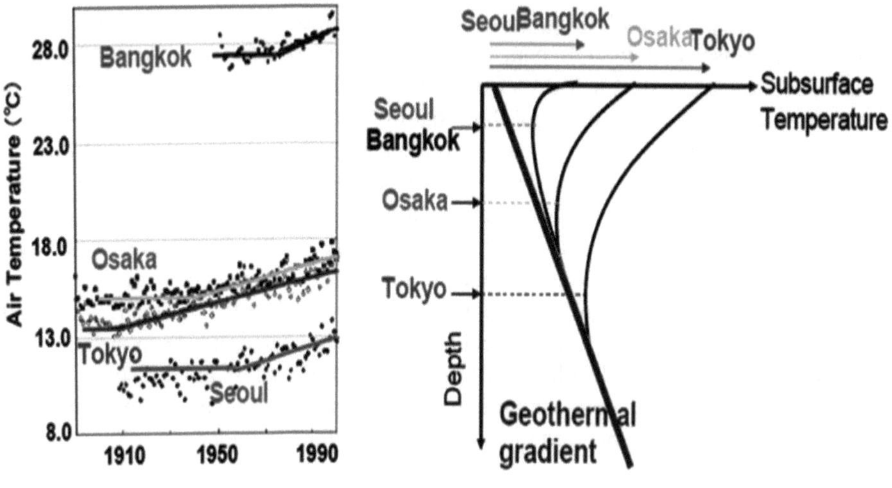

Fig. 4.4 Effects of surface warming due to global warming and the heat island effect on subsurface temperature

The heat island effect on subsurface temperature is an important global groundwater quality issue, because it may alter groundwater systems geochemically and microbiologically (Knorr et al. 2005). Many cities around the world, including in Asia, are experiencing rapid population increase. Reconstructions of surface warming history using subsurface temperature have been made in rural, suburban, and urban areas of Bangkok (Yamano et al. 2009). Surface warming started earlier in the current urban area, followed by the current suburban areas, and then rural areas. Therefore, the heat island effect reflects the history of city expansion.

Subsurface temperature usually increases with depth because of geothermal energy from deep inside the Earth, but surface warming due to global warming and urbanization changes this situation. A new thermal boundary exists not at the surface but deeper in the subsurface, such as 100 m below the surface in the case of Tokyo.

4.5 Changes in Land Cover/Use

Land cover/use is the key factor in partitioning water, dissolved material and heat energy into surface and subsurface environments. The results of expanding urban area (houses and industries) in Tokyo, Osaka, and Seoul are shown in Fig. 4.5. The Tokyo urban area expanded between 1930 and 1970 by 788 km^2 and between 1970 and 2000 by 2,991 km^2, in Osaka by 569 and 907 km^2, and in Seoul by 196 and 954 km^2, respectively. In general, the decrease of permeability associated with increased urban extent occurs later, followed by a subsequent reduction of the groundwater recharge rate. The increase in urban area also causes an increase in aquifer thermal storage, which is shown by the magnitude of the heat island effect (Taniguchi et al. 2009d).

Analyses of land-cover/use changes were conducted for Tokyo, Osaka, and Seoul with nine categories including, forests, housing, industries, paddy fields, other agriculture, grass/wasteland, ocean, water and wet land. The magnitude of change in urban area was greatest in Tokyo from 1970 to 2000. The magnitude of change in urban area was also larger in Seoul than that in Osaka, from 1970 to 2000.

Because of the reduction of the groundwater recharge rate, the reliability of groundwater as a resource may decrease after urbanization, and groundwater contamination, including subsurface thermal anomalies, may increase. However, a novel subsurface environmental problem has occurred in Tokyo. The city has experienced a floating subway station due to buoyancy associated with recovering groundwater potential after regulation of pumping was introduced. This is attributed to sufficient recovered groundwater recharge in cities located in monsoon Asia. The development of integrated indicators based on GIS is necessary for better understanding the relationship between human activities and the subsurface environment, and proper management beyond the surface–subsurface boundary.

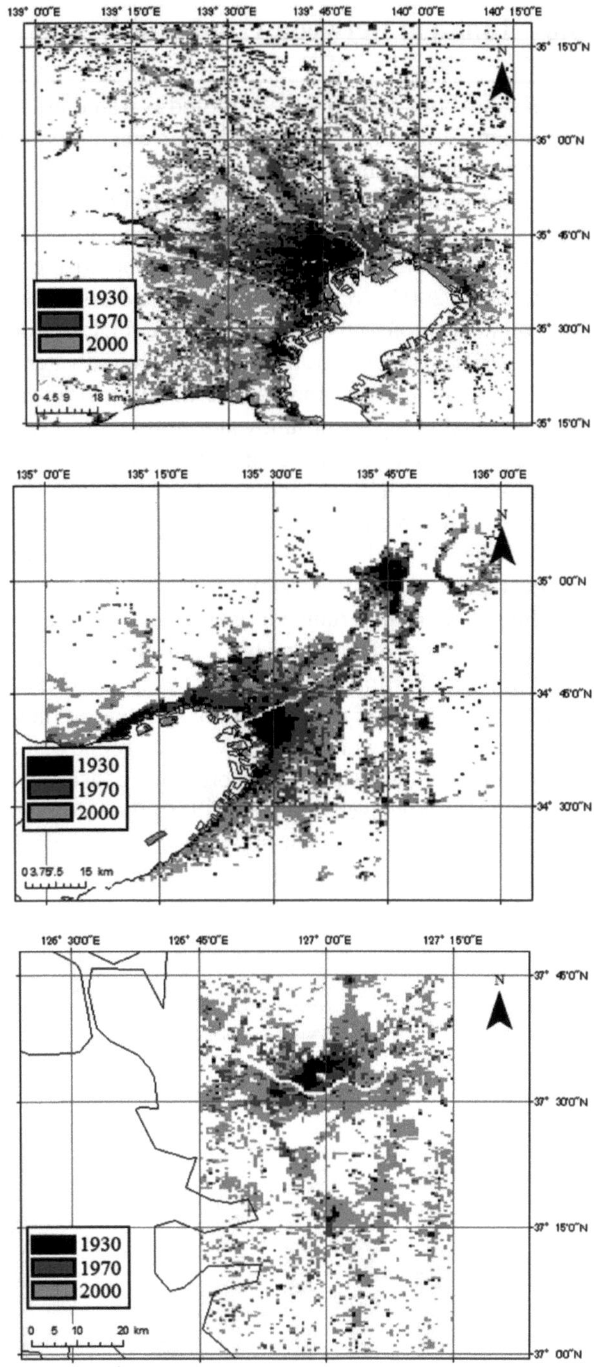

Fig. 4.5 Changes in urban area in Tokyo (*top*), Osaka (*middle*), and Seoul (*bottom*)

4.6 Linkage Between Land and Ocean

The second boundary linkage explored is water and material transport between land and ocean. Soil water and groundwater contamination are important aspects of the subsurface environment, because accumulated subsurface contaminants eventually discharge as material (contaminant) transport to the rivers and coast. Recent studies have shown that direct groundwater discharge to the coastal zone is a significant water and material pathway from the land to the ocean (Taniguchi et al. 2002; Burnett et al. 2003). Many water quality and associated problems influencing coastal environments around the world today are related to past and ongoing contamination of terrestrial groundwater, because that groundwater is now seeping out along many shorelines. For instance, inputs of fertilizers and sewage on land over several decades have resulted in higher groundwater nitrogen levels, and its slow yet persistent discharge along the coast may eventually result in coastal marine eutrophication. Such inputs may contribute to increased occurrences of coastal ecosystem problems. Since most Asian cities are coastal, material and contaminant transport by groundwater is key to understanding present and future coastal water pollution and its effects on associated ecosystems.

The potential importance of groundwater seepage to nutrient inputs into Manila Bay and The Bay of Thailand has been established by a multidisciplinary approach utilized in the Philippines (Taniguchi et al. 2007a) and Bangkok (Fig. 4.6, Burnett et al. 2007). Seepage meters, resistivity measurements, and use of natural Rn as a groundwater tracer were used to evaluate groundwater discharge into coastal waters. Seepage meter and tracer results provided consistent results of estimates of submarine groundwater discharge (SGD) into the two bays. Both methods also showed that seepage fluxes are not steady-state but are modulated by tides. Resistivity profiles showed that the saline–freshwater interface moves on a tidal time scale.

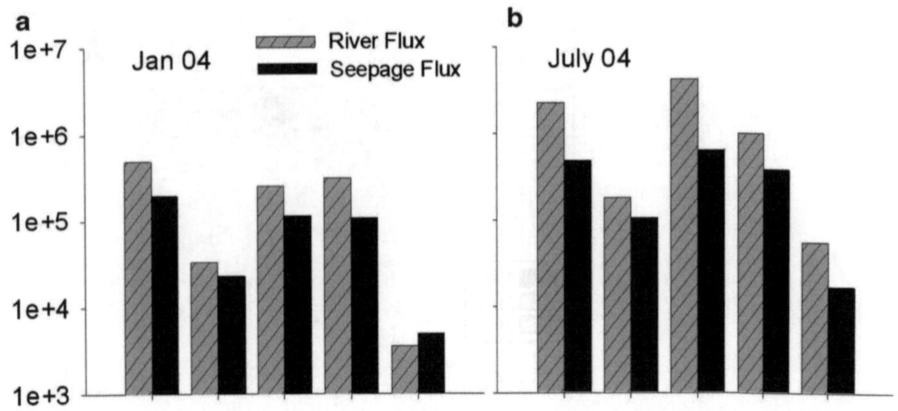

Fig. 4.6 Material transport by groundwater (seepage flux) and river water (flux) during the (**a**) dry and (**b**) wet seasons at the Bay of Thailand (modified from Burnett et al. 2007)

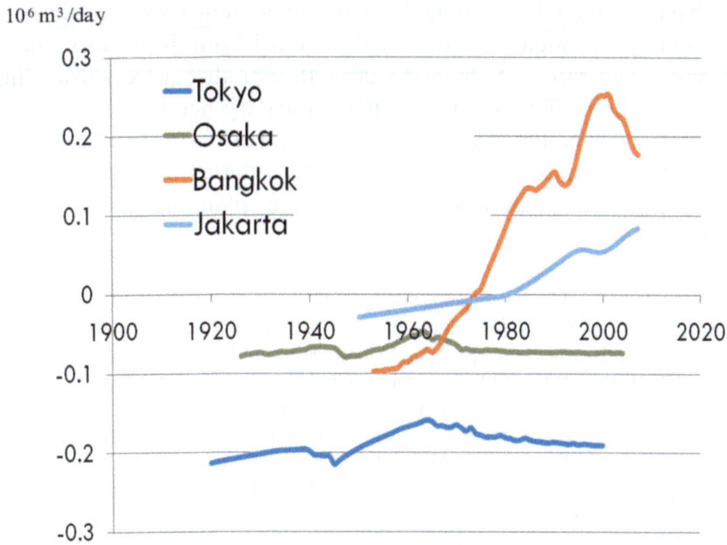

Fig. 4.7 Direction and magnitude of water exchange between land and ocean

Our results showed that dissolved inorganic nitrogen (DIN) fluxes via SGD are comparable in magnitude to DIN fluxes from each of the two major rivers that drain, respectively, into Manila Bay (Taniguchi et al. 2007a), and Thai Bay (Burnett et al. 2007; Taniguchi et al. 2009c).

Another example of SGD offshore was found in the Chokai Mountain area, northern Japan. Use of Rn to detect SGD along the study transect perpendicular to the coast of Mega Bay showed that SGD occurred 1,000 m offshore. Therefore, a new groundwater boundary between land and ocean may exist 1 km offshore.

Material is also transported from the ocean to the land. Changes in the direction and magnitude of the water flux between land and ocean are shown in Fig. 4.7. Flux calculations were made based on MODFLOW simulations for Tokyo, Osaka, Bangkok and Jakarta. Detailed methods are presented by Taniguchi and colleagues (2010). The net water flow direction in Tokyo and Osaka was from land to ocean, although saltwater intrusions occurred in some coastal areas (Fig. 4.7). The direction of water flow in Bangkok and Jakarta changed from land to ocean around 1970, and then from ocean to land around 1980, because of groundwater pumping after urbanization.

4.7 Conclusions

Linkages between surface and subsurface environments and between the land and ocean were evaluated in the RIHN project "Human impacts on urban subsurface environments". Evaluations of water, heat, and material transfers beyond the surface–subsurface boundary and the land–ocean boundary were conducted. A new thermal boundary was identified moving downward from the surface due to global warming and urbanization. Submarine groundwater discharge shows impacts of land water extending offshore, and the new boundary needs to be used for better environmental management instead of the traditional basin. Evaluations of water, material and thermal transport beyond these boundaries are important for water resource management, coastal ecology, and urban planning in the Asian coastal mega-cities.

Acknowledgments I would like to acknowledge all members of the USE (Urban Subsurface Environment) project of RIHN (Research Institute for Humanity and Nature) for helping in conducting this research project; in particular, Robert Delinom (LIPI, Indonesia), Somkid Buapeng (Ministry of Natural Resources and Environment, Thailand), Fernando Siringan (University of the Philippines), Chung-Ho Wang (Academia Sinica, Taiwan), and Backjin Lee (Korea Research Institute for Human Settlements) for their great support in the field. This research is closely linked with the UNESCO-GRAPHIC project (Groundwater Resources Assessment under the Pressures of Humanity and Climate Changes, Aureli and Taniguchi 2006). This study is also linked with international organizations including IAHS, IGRAC, IAH, GWSP, IGBP-LOICZ, and IHDP. I thank those organizations for their useful collaboration.

References

Aureli A, Taniguchi M (2006) Groundwater assessment under the pressures of humanity and climate changes. UNESCO-IHP brochure. UNESCO, Paris

Burnett WC, Bokuniewicz H, Huettel M, Moore WS, Taniguchi M (2003) Groundwater and pore water inputs to the coastal zone. Biogeochemistry 66:3–33

Burnett WC, Wattayakorn G, Taniguchi M, Dulaiova H, Sojisuporn P, Rungsupa S, Ishitobi T (2007) Groundwater-derived nutrient inputs to the upper Gulf of Thailand. Cont Shelf Res 27:176–190

Huang S, Pollack HN, Shen Po-Yu (2000) Temperature trends over the past five centuries reconstructed from borehole temperatures. Nature 403:756–758

Knorr W, Prentice IC, House JI, Holland EA (2005) Long-term sensitivity of soil carbon turnover to warming. Nature 433:298–301

Lieberman V (2003) Strange parallels: southeast Asia in a global context, c.800–1830. Cambridge University Press, New York

Ness GD, Prem T (eds) (2004) Asian urbanization for the new millennium. Michael Cavendish, Singapore

Taniguchi M, Uemura T (2005) Effects of urbanization and groundwater flow on the subsurface temperature in Osaka, Japan. Phys Earth Planet In 152:305–313

Taniguchi M, Burnett WC, Cable JE, Turner JV (2002) Investigation of submarine groundwater discharge. Hydrol Process 16:2115–2129

Taniguchi M, Burnett WC, Dulaiova H, Siringan F, Foronda J, Wattayakorn G, Rungsupa S, Kontar EA, Ishitobi T (2007a) Groundwater discharge as an important land–sea pathway into Manila Bay, Philippines. J Coast Res 24(1A):15–24

Taniguchi M, Uemura T, Jago-on K (2007b) Combined effects of urbanization and global warming on subsurface temperature in four Asian cities. Vadose Zone J 6:591–596

Taniguchi M, Burnett WC, Ness GD (2009a) Integrated research on subsurface environments in Asian urban areas. Sci Total Environ 407:3076–3088

Taniguchi M, Shimada J, Fukuda Y, Yamano M, Onodera S, Kaneko S, Yoshikoshi A (2009b) Anthropogenic effects on the subsurface thermal and groundwater environments in Osaka, Japan and Bangkok, Thailand. Sci Total Environ 407:3153–30164

Taniguchi M, Shimada J, Fukuda Y, Yamano M, Onodera S, Buapeng S, Delinom R, Siringan F, Wang CH, Lee B, Yasumoto J, Yamamoto K (2009c) Degradation of subsurface environment due to human activities and climate variability in Asian cities. IAHS Publication 329: 124–129

Taniguchi M, Yoshikoshi A, Kaneko S, Buapeng S, Delinom R, Siringan F, Wang CH, Lee Shiraki Y, Endo, T, Yamashita A (2009d) Changes in the reliance on groundwater versus surface water resources in Asian cities. IAHS Publication 330:218–224

Umezawa Y, Hosono T, Onodera S, Siringan F, Buapeng S, Delinom R, Jago-on KA, Yoshimizu C, Tayasu I, Nagata T, Taniguchi M (2009) The characteristics of nitrate contamination in groundwater at developing Asian-Mega cities, estimated by nitrate $d^{15}N$ and $d^{18}O$ values. Sci Total Environ 407:3219–3231

Yamano M, Goto S, Miyakoshi A, Hamamoto H, Lubis RF, Monyrath V, Kamioka S, Huang S, Taniguchi M (2009) Study of the thermal environment evolution in urban areas based on underground temperature distributions. Sci Total Environ 407:3120–3128

Chapter 5
The Nitrate–Arsenic Boundary as an Important Concept in Aquatic Environmental Studies

Takahiro Hosono

Abstract The nitrate and arsenic (NA) boundary is proposed as an important new boundary concept for catchments. It is defined as the redox border distinguishing whether nitrate or arsenic can be present in water. The NA boundary concept is explained based on the role of the redox system and by introducing research examples which use nitrate and sulfate isotope ratios from urban catchments in a variety of Asian countries. The global-scale importance of the NA boundary concept for sustainable groundwater use is illustrated based on a compiled Asian dataset.

Keywords Asia • Boundary • Groundwater • Pollution • Redox

5.1 Definition of the NA Boundary

Redox (reduction/oxidation) potential is an important factor controlling the solubility of ions and elements in water at surface and subsurface environments. Redox processes affect the presence of potentially toxic ions or elements such as nitrate (NO_3^-) and arsenic (As) in water. Therefore, determining the kinds of redox processes present and documenting their spatial distributions are essential for understanding the status of water quality and the formation mechanisms of pollutants in catchment areas.

In natural systems, the variations in redox potential, and thus water quality, are controlled by biogeochemical reactions through microorganism activity. Microorganisms that catalyze redox processes in natural systems compete for limited resources (Fig. 5.1). For example, dissolved oxygen (O_2) is the most

T. Hosono (✉)
Priority Organization for Innovation and Excellence, Kumamoto University,
2-39-1 Kurokami, Kumamoto 860-8555, Japan
e-mail: hosono@kumamoto-u.ac.jp

M. Taniguchi and T. Shiraiwa (eds.), *The Dilemma of Boundaries: Toward a New Concept of Catchment*, Global Environmental Studies,
DOI 10.1007/978-4-431-54035-9_5,

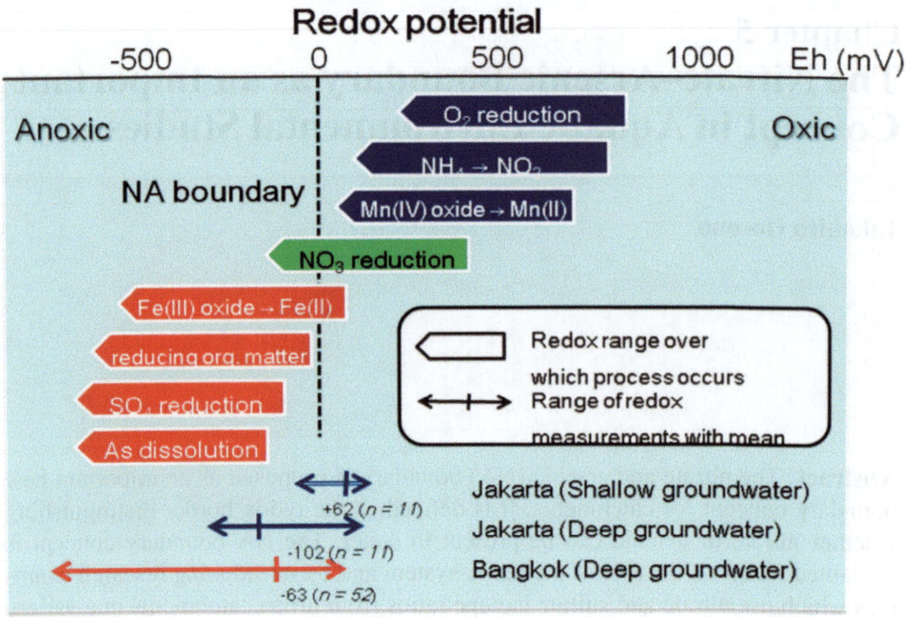

Fig. 5.1 Sequence of microbially mediated redox processes (modified after UNEP 2003)

preferred electron accepter used by microorganisms. Under anoxic conditions, nitrate is the next most energetically favorable naturally available electron acceptor for nitrate reducing bacteria, followed, for example, by sulfate (SO_4^{2-}) which is reduced through the activity of sulfate reducing bacteria (Fig. 5.1). Likewise, the mobilization of metal elements changes (Mn (IV)→Mn (II) and then Fe (III)→Fe (II)) as reducing processes proceed.

Nitrate and arsenic are the most common and important polluting ions or elements of concern. Importantly, the redox potential (E_h) range shows a big gap between where nitrate reduction is complete and dissolution of arsenic occurs (Fig. 5.1). Nitrogen in the nitrate form can exist under oxic conditions with a redox potential over 0 mV. However, it cannot be present in this form at a redox potential near 0 mV because denitrification occurs immediately under anoxic conditions. On the contrary, arsenic can be dissolved under anoxic conditions with a redox potential under 0 mV, such as in an environment where sulfate reduction occurs. From the above explanation, it can be seen that, in terms of nitrate and arsenic pollution, redox conditions can be divided into two categories (1) $E_h > 0$ mV in a range where nitrogen can be present as a form of nitrate ion but dissolved arsenic cannot be present and (2) $E_h < 0$ mV in a range where nitrate has been denitrified but arsenic can be dissolved in the water. It is still not clear if the border between the above two different redox conditions can be placed at $E_h = 0$ mV. Here, the boundary between these two different redox conditions which probably occurs at E_h values around 0 mV is defined as the nitrate and arsenic (NA) boundary.

The purpose of this work is to propose a new important boundary concept for water catchments, beyond a simple boundary between the surface and subsurface. Compared with surface water, subsurface water is commonly anaerobic, i.e. situated in reducing conditions. However, some unconfined aquifers could have aerobic subsurface conditions, where parts of the groundwater can oxidize, with a redox range similar to that of the surface environment. Therefore, an understanding of the reduction/oxidation boundary in subsurface environments is important for understanding the formation mechanisms that affect water quality beyond the surface–subsurface boundary. In particular, the NA boundary is one of the most important redox boundaries in terms of water quality. In this article, the focus is placed on the relationship between nitrate and arsenic concentrations in Asian metropolitan aquifers, to emphasize the importance of the NA boundary based on both the author's published (Hosono et al. 2009a, b, c, d; 2010; 2011a, b, c; Umezawa et al. 2009) and unpublished data.

In the next section (Sect. 5.2), the nitrate and arsenic pollution mechanism is explained in relation to redox processes. In Sect. 5.3, examples of NA boundaries are shown in case studies of Jakarta and Bangkok aquifers, which are situated in contrasting settings in terms of redox environments. Nitrate and sulfate isotope ratios are useful for demonstrating the occurrence of reducing processes from bacterial activities. Applications of these isotope methods are also described in this section. In Sect. 5.4, nitrate and arsenic data from Asian metropolitan aquifers are compiled to review the water quality of these areas and discuss the importance of the NA boundary for sustainable groundwater management in Asian countries.

5.2 Pollution and Attenuation Mechanisms Related to Redox Change

Mechanisms for the generation of nitrate ions and dissolution of arsenic due to a change in aerobic/anaerobic conditions are shown in Figs. 5.2 and 5.3, respectively. In general, dissolved nitrate ions are formed from ammonium nitrogen from both natural and anthropogenic origins through the nitrification process under relatively aerobic conditions (Fig. 5.2). Nitrate from atmospheric and/or nitrate fertilizer origins can also be present under aerobic conditions. Under aerobic conditions, nitrate pollution will occur when there are significant nitrate inputs. On the other hand, the denitrification process will proceed if nitrate ions are placed under anaerobic conditions to form nitrite, and ultimately, nitrogen gas (Fig. 5.2). Therefore, it can be seen that nitrogen pollution and attenuation result from a balance between aerobic or anaerobic conditions. According to Fig. 5.1, the denitrification process might be completed at a redox potential around 0 mV.

In contrast to nitrate, arsenic is believed to be dissolved in water under anaerobic conditions in many cases, except in areas where mines and man-made pollution are

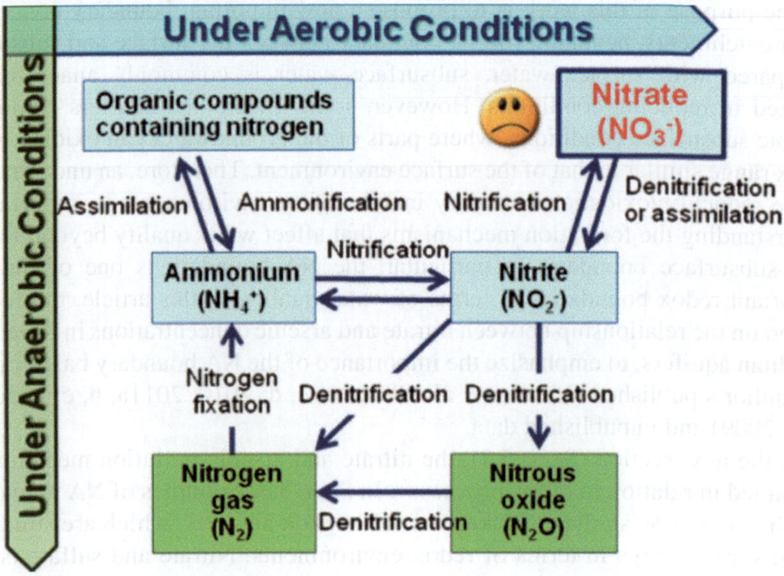

Fig. 5.2 Transformation of nitrogen from waste water during groundwater recharge (modified from Lawrence et al. 1997)

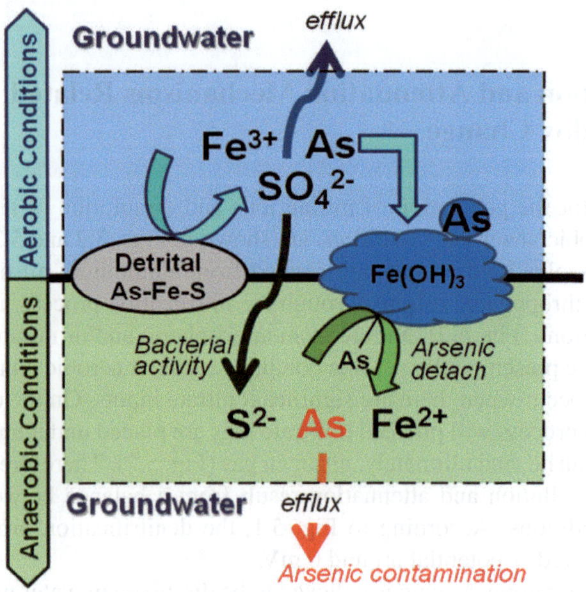

Fig. 5.3 Transformation of arsenic, sulfur, and iron in the natural systems

responsible for its distribution (Fig. 5.3). Naturally-occurring arsenic, once absorbed on the surface of ferric hydroxide under aerobic conditions, can be dissolved in water if it moves to anaerobic conditions, hence becoming a source of pollution (there are other mechanisms by which arsenic can dissolve in water through the oxidation process, but these are not discussed here). Accompanying the arsenic dissolution, reduction of Fe(II) from Fe(III) oxide and reduction of sulfate occur through bacterial activities (Fig. 5.3). Therefore, the species of iron present and the occurrence of sulfate reduction could be good parameters for monitoring arsenic pollution. According to Fig. 5.1, the above dissolution processes might occur at redox potentials far below 0 mV. This also suggests that areas polluted with arsenic could be attenuated if the aquifer environments become more oxic.

The vertical distributions of redox boundaries are commonly controlled by the constituent materials and the structure of the aquifer geology under the subsurface environment. It is reasonable to think that aerobic conditions are developed in crystalline rock aquifers with joints and structural deformations, where interactions with surface waters are active through the structural breaks. In contrast, anaerobic conditions would be dominant in aquifers covered by impermeable clay layers where it is difficult for interactions between surface and subsurface materials to take place, even at shallow depths. In volcanic fan areas that usually have many springs and a lot of groundwater, aerobic conditions are commonly developed in shallow unconfined aquifers but anaerobic conditions can be observed in deep confined aquifers.

In addition to geological controls, groundwater pumping is an important factor in changing redox conditions in subsurface environments. Groundwater pumping from an aquifer with reducing conditions might enhance interaction with surface waters, making redox conditions in this aquifer more oxic. Such redox changes increase the risk of nitrate pollution when there are substantial amounts of ammonium ions present in the aquifer. The other risk is arsenic pollution, as dissolution occurs if reduction of an aquifer environment takes place. It has been said that the number of people suffering from arsenic pollution increased with the increase in the number of wells available for everyday use in areas such as Bangladesh and West Bengal. Groundwater pumping can be an unwanted cause of secondary water pollution.

The Asian metropolitan cities of Jakarta and Bangkok are on fan deltas near coastal zones. Of the many similarities and differences between these two areas, the most important point to be mentioned here is the difference in the geological materials covering the land surfaces. Briefly, the uppermost section of the Jakarta fan area is composed of permeable volcanic deposits, whereas that of the Bangkok area (lower central plain) is covered by thick impermeable clay layers. This contrast in geological setting must create different features in their redox environments. In the next section, the influences of geology and groundwater pumping on the redox conditions (NA boundary) and the groundwater quality (nitrate and arsenic pollution) are described by using Jakarta and Bangkok as a case study.

5.3 Case Study of Jakarta and Bangkok

5.3.1 Location, Geology, Hydrogeology and Land Use

Jakarta The Jakarta area is located in the northwest part of the Java Islands, Indonesia. The geology of the Jakarta area is composed mainly of sedimentary basement rocks of the Tertiary age overlying Pleistocene sedimentary deposits, Quaternary volcanic rocks and their deposits, making a typical volcanic fan topography, and alluvium (Figs. 5.4 and 5.5). Two types of aquifers are situated in the Quaternary volcanic deposits and Pleistocene sedimentary deposits (Fig. 5.5): (1) the first aquifer (unconfined aquifer) is approximately 80 m from the surface level and has relatively aerobic conditions (Fig. 5.1) and (2) the second aquifer (confined aquifer) is about 100–250 m from the surface level and has relatively anaerobic conditions. Hereafter we refer to the groundwater from the first aquifer as shallow groundwater and that from the second aquifer as deep groundwater.

The Jakarta metropolitan area, including the suburban urbanized area, is one of the largest megalopolises in Asia, with a population of around 13.6 million. The highly populated urbanized area (including industrial and commercial areas) is distributed in the northern plain and coast, whereas the agricultural area with fruit farms and orchards is distributed in the middle of the volcanic alluvium fan (Fig. 5.6).

Bangkok The Lower Central Plain (LCP), a huge Quaternary deltaic plain (25,000 km^2), is situated in the central part of Thailand, surrounded by a mountainous range with an altitude of >1,000 m. A large amount of fluvial sediments, which are deposited in the Bangkok Basin or Thon Buri Basin, form the LCP with its total thickness of 2,000 m maximum (Sinsakul 2000). The geology of the LCP is composed mainly of sedimentary basement rocks from the Tertiary age, alternating layers of fine to coarse sand of the Pleistocene age and impermeable consolidated clay ca. 30 m thick of the Holocene age—so called Bangkok clay—(Figs. 5.4 and 5.5). Aquifers are developed in the alternating layers of fine to coarse sand of the Pleistocene age (Fig. 5.5). Stratigraphically, the upper 600 m of sediments are divided into eight aquifers: the Bangkok (BK: 16–110 m below ground level), Phra Pradaeng (PD: 60–130 m), Nakhon Luang (NL: 100–210 m), Nonthaburi (NB: 170–270 m), Sam Khok (SK: 240–330 m), Phya Thai (PT: 276–360 m), Thon Buri (TB: 350–500 m) and Pak Nam (PN: 420–500 m) aquifers. Although they are stratigraphically classified as above, aquitards between them are not always continuous.

In the LCP, the Bangkok metropolis (population of approximately 6.5 million in an area of 1,010 km^2) is located about 25 km to the north of the Gulf of Thailand on the flood plain of the Chao Phraya River (Fig. 5.6). Most of the flood plain behind the urbanized area is occupied by paddies maintained by irrigation. Orchard fields are distributed to the west of the urbanized area.

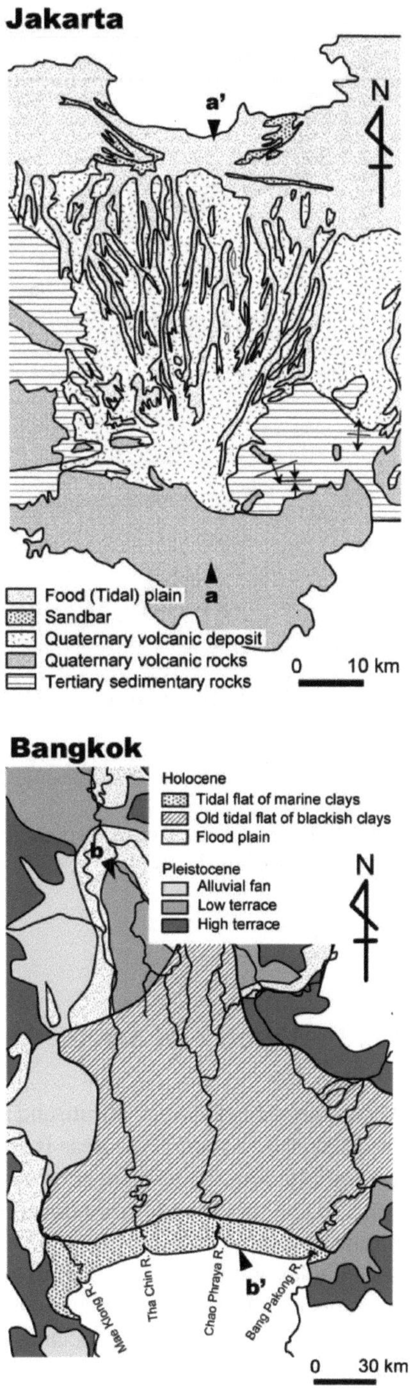

Fig. 5.4 Simplified geological maps for the Jakarta area (modified from Sidarto et al. 1992) and the LCP (Bangkok City) areas (modified from Sinsakul 2000)

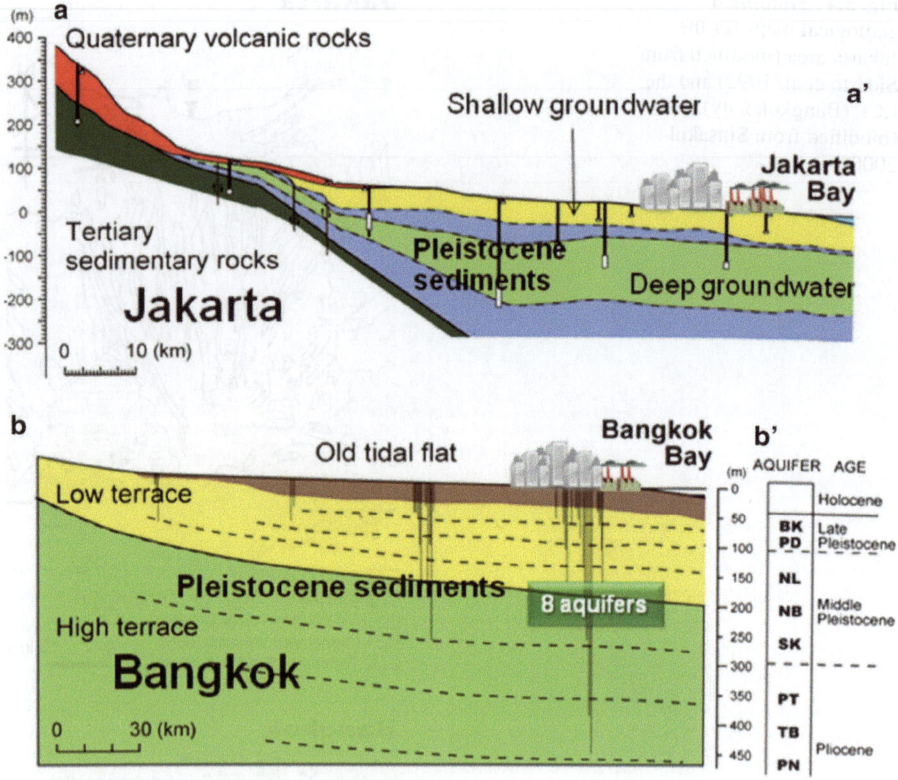

Fig. 5.5 Hydrogeological cross sections of the Jakarta and LCP areas. The sections a–a′ and b–b′ are the same as those in Fig. 5.4

5.3.2 Nitrate and Arsenic Pollution in Jakarta and Bangkok

The relationship between nitrate and arsenic concentrations of groundwater samples from Jakarta and Bangkok areas is shown in Fig. 5.7. The figure shows that, in general, nitrate ions are present in groundwater when arsenic is not present, and vice versa. It can be assumed from the above that different redox conditions make either nitrate or arsenic available in the water. For the Jakarta area, the shallow groundwater samples with relatively higher redox potentials (Fig. 5.1) contain significant amounts of nitrate but no arsenic, while deep groundwater samples with relatively lower redox potentials are concentrated in arsenic but not nitrate. Onthe other hand, the groundwater samples from the LCP area, which has a relatively lower redox potential (Fig. 5.1), contain almost no nitrate but have significant arsenic concentrations.

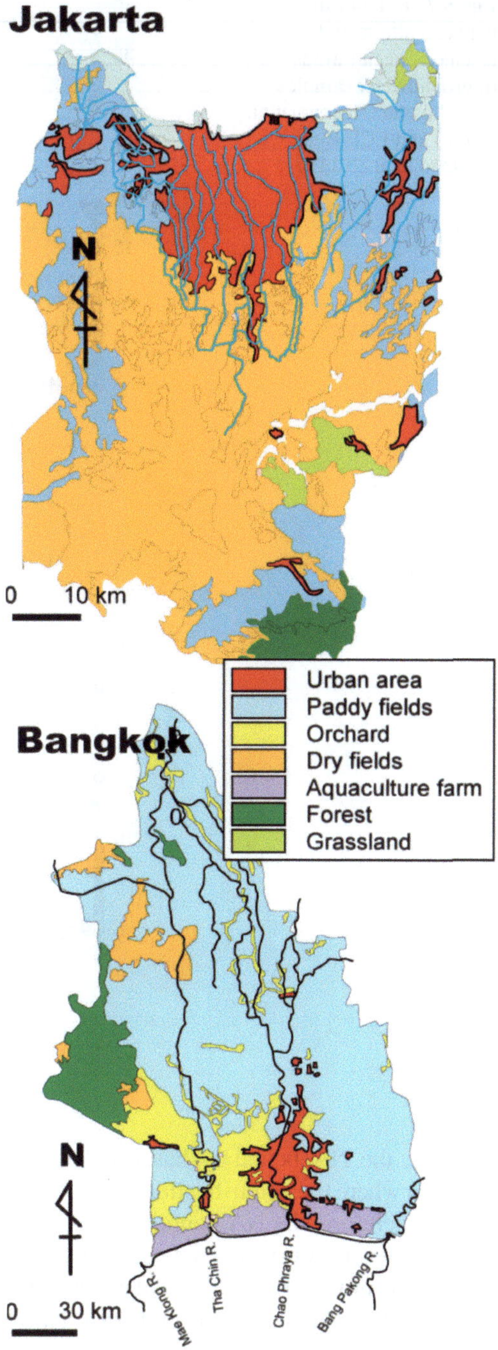

Fig. 5.6 Land-use maps for Jakarta (data supplied by the Indonesian Institute of Science) and LCP (data supplied by the Ministry of Natural Resources and Environment in Thailand) areas

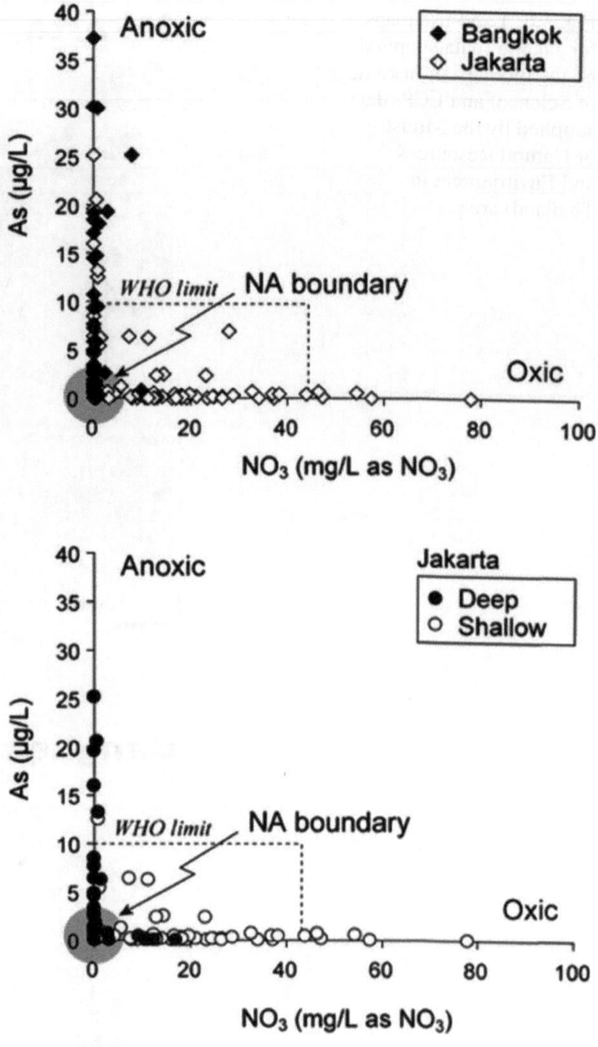

Fig. 5.7 Relationship between nitrate and arsenic concentrations of groundwater samples from Jakarta (Hosono et al. 2011c) and LCP areas (Hosono et al. 2011b)

Here, the NA boundary has been defined as the boundary of redox potential between where the denitrification process is complete and where arsenic starts to dissolve. Therefore, the NA boundary is placed in Fig. 5.7 at the origin of the coordinates. We can consider, then, that the NA boundary for Jakarta groundwater is distributed in the boundary between shallow and deep groundwaters. On the other hand, the NA boundary in the LCP area is distributed in the Bangkok clay layer above the Bangkok Aquifer. These results imply that the difference in the geological setting for each area controls the position of the NA boundary, and furthermore, the kinds of dissolved ions or elements and types of pollution present in the groundwater.

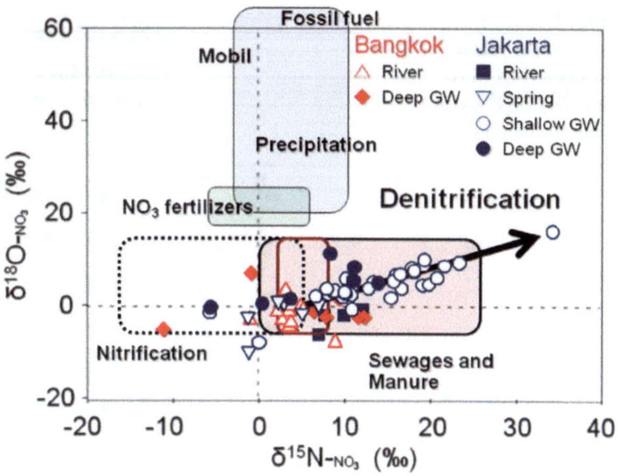

Fig. 5.8 $\delta^{15}N$ vs. $\delta^{18}O$ of nitrate for water samples collected from Jakarta and LCP areas (data from Umezawa et al. 2009)

The water chemistry in both areas suggests the occurrence of denitrification and sulfate reduction via bacterial activities. However, the concentration data themselves are not direct evidence of the occurrence of bacterial activity. In the next part, research results using nitrate and sulfate isotope ratios are introduced, as these ratios are powerful tools for tracing the participation of bacteria in systems.

5.3.3 *Nitrate and Sulfate Isotope Ratios*

In general, bacteria preferentially use lighter isotopes when they create energy from nitrate and sulfate ions. As a result, the remaining waters become dominant in heavier isotopes (higher isotope ratios) due to isotope fractionation. This enables nitrate and sulfate isotope ratios to be used in determining the occurrence of bacterial activities in addition to elucidating the sources of dissolved nitrate and sulfate. The principles and applications of isotope methods are described in numerous texts (Kendall 1998; Cook and Herczeg 2000; Hoefs 2004). Here, emphasis is placed on the results from the Jakarta and LCP areas to show the isotopic evidence for the occurrence of bacterial activity (Figs. 5.8 and 5.9). Three important points were suggested by the nitrate and sulfate isotopic studies on the Jakarta and LCP areas, as follows (Figs. 5.8 and 5.9; Umezawa et al. 2009; Hosono et al. 2011c):

1. On the nitrate isotope diagram (Fig. 5.8), almost all the groundwater samples are plotted on the compositional trend of denitrification, suggesting the occurrence of bacterial activity in the subsurface environments in both Jakarta and the LCP area. Although significant nitrogen inputs were assumed in the Jakarta shallow groundwater, nitrogen pollution was reduced through the denitrification process.

Fig. 5.9 $\delta^{34}S$ vs. $\delta^{18}O$ of sulfate for water samples collected from (**a**) Jakarta and (**b**) LCP areas (unpublished data)

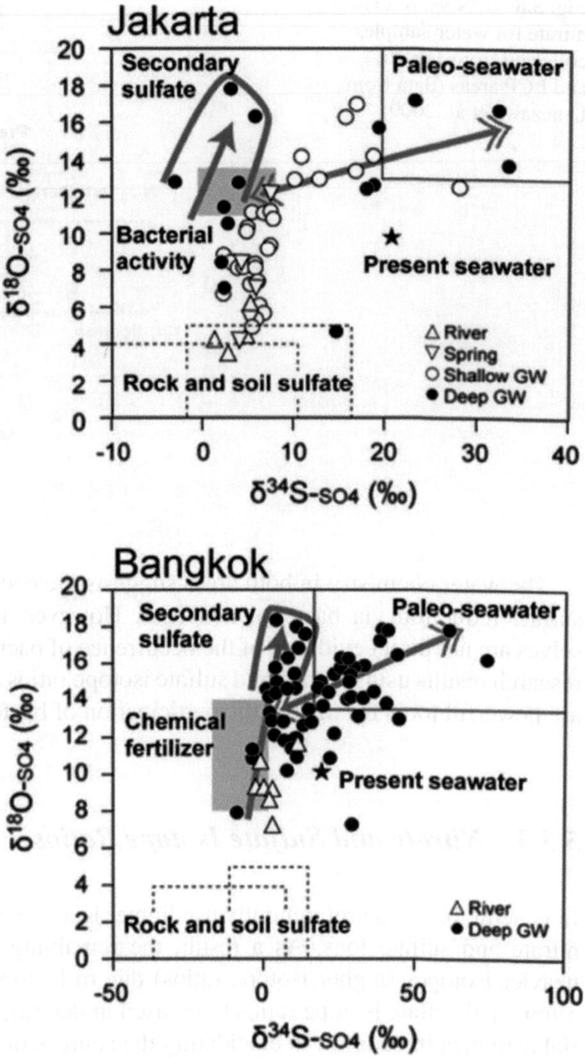

2. On the sulfate isotope diagram (Fig. 5.9), almost all the groundwater samples from the LCP area and the Jakarta deep aquifer are plotted on the compositional field of water that is affected by the sulfate reducing process, whereas those from the Jakarta shallow aquifer are plotted in other compositional fields. These results correspond with the observation that the Jakarta shallow aquifers are dominated by nitrate ions but arsenic can be dissolved in the deep Jakarta groundwater and the LCP aquifer.
3. On the sulfate isotope diagram (Fig. 5.9), some groundwater samples from both areas are plotted on the compositional field of secondary sulfate, suggesting that, under reducing conditions, sulfate ions are generated by the oxidation of natural sulfur of geological origin by sulfur oxidizing bacteria.

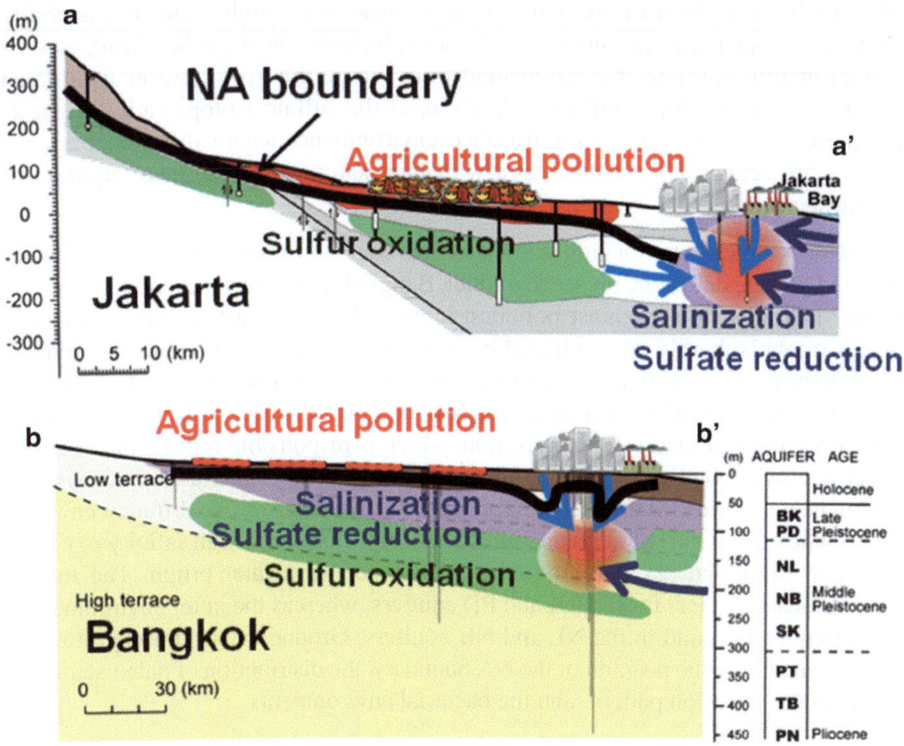

Fig. 5.10 Cross sections illustrating the hydro-bio-geochemical processes in the Jakarta and LCP areas

From the above, it can be seen that the presence of bacterial activity, its type, and the effects on water quality can be evidenced by isotope ratios. A comparison of isotope data from aquifers in different areas (countries) could characterize the bio-geochemical features in each area. The hydro-bio-geochemical processes for the Jakarta and LCP areas, based on the distribution of elemental and isotope data, are shown in Fig. 5.10.

5.3.4 *Hydro-Biogeochemical Models for Jakarta and Bangkok*

The positions of the NA boundaries for Jakarta and the LCP area are clearly different (Fig. 5.10). The biogeochemical features and the status of the nitrate and arsenic pollution accompanying the position of the NA boundary are summarized for the Jakarta and LCP areas as follows:

Jakarta Agricultural activity significantly affects the shallow groundwater contamination. Nitrate pollution is typically found in an area composed of volcanic materials.

The aerobic condition of the volcanic host rocks with high water permeability seems to be related to the presence of contamination. On the other hand, arsenic contamination is found in deep groundwater where the groundwater is situated in relatively anaerobic conditions. Analysis of the sulfate isotope ratios suggests the presence of two different subsurface environments where different bacterial species are active: (1) a coastal area with high salinity and (2) deep aquifers of freshwater origin.

Bangkok Nitrate contamination is not a serious concern for most of the groundwater in the LCP area, due to the presence of thick Bangkok clay which acts both as a redox barrier and as a shield against pollution inputs. However, surface waters and soil waters might be highly contaminated by agricultural activities. In general, attention needs to be paid to naturally occurring arsenic, because groundwaters with anaerobic conditions are situated all over the aquifers. In addition, the Bangkok clay with paleo-seawater causes groundwater salinization, which is responsible for one of the major water quality problems in the LCP area. As shown for the Jakarta area, analysis of sulfate isotope ratios for the LCP area suggests the presence of two different environments where bacterial sulfate reduction occurs: (1) aquifers of high salinity of paleo-seawater origin and (2) aquifers of low salinity of freshwater origin. The former environment is typical for the BK and PD aquifers, whereas the latter environment is most frequently found in the NL and NB aquifers. Groundwater pumping in urban areas is disturbing the position of the NA boundary, the distribution of paleo-seawater, the seawater intrusion pattern, and the bacterial environments.

5.4 Application of NA Boundary for Sustanable Groundwater Use

The relationships between the nitrate and arsenic concentrations of groundwaters from Asian metropolitan aquifers (Seoul, Taipei, Bangkok, Manila and Jakarta) are shown in Fig. 5.11. The groundwater samples from the Taipei Basin were taken from unconfined aquifers with depths less than 20 m from the surface. The Seoul groundwaters were collected from crystalline-hosted aquifers with depths of 10–80 m, and the Manila groundwaters were from both unconfined and confined aquifers with depths of a few meters to 300 m.

Taipei shallow groundwater, which has an aerobic environment, displayed similar concentration features to Jakarta shallow groundwater; that is it had a high nitrate concentration relative to that of arsenic (Fig. 5.11) (there were some shallow groundwaters with high arsenic concentrations, but these are not shown here). Seoul groundwater samples showed similar nitrate and arsenic concentration characteristics, even for the samples collected from a depth of 80 m (Hosono et al. 2009c). In contrast, Manila groundwater samples displayed low nitrate but high arsenic concentrations for both shallow and deep aquifers (Fig. 5.11). It can be assumed that this means reducing environments are developing under this area in the same way as for the LCP area.

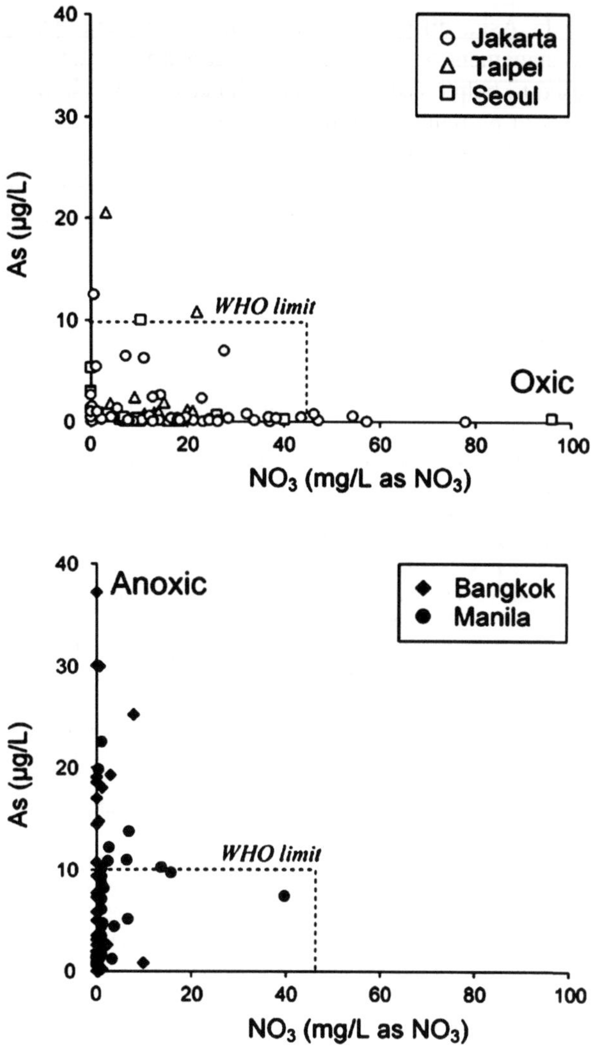

Fig. 5.11 Relationships between nitrate and arsenic concentrations of groundwater samples from Asian metropolitan aquifers. Data sources: Seoul (Hosono et al. 2009c), Manila (Hosono et al. 2010), Taipei (Hosono et al. 2011a), Bangkok (Hosono et al. 2011b), and Jakarta (Hosono et al. 2011c)

Consequently, it is shown that water quality in metropolitan aquifers can be categorized based on the position of the NA boundary. On the basis of this boundary, Jakarta and Taipei shallow aquifers and Seoul crystalline rock aquifers are areas where nitrate can dissolve but little arsenic is present. In contrast, LCP and Manila aquifers are situated at the position where arsenic can be present but nitrogen ions cannot be present in the form of nitrate. It is clear from the above conclusions that the NA boundary is a practical and applicable concept for water resource preservation and management in Asian countries. With a similar aim, compilation of a redox distribution map based on an assemblage of several redox indicators has been attempted for the United States (USGS 2009).

In Asia, different countries have different strategies for the preservation and management of water resources. It is hoped that the NA boundary concept will be used to help construct a water management system that will reach beyond the borders of particular countries in Asia, as arsenic pollution is a major problem in this area. However, to date not enough data has been compiled. Further sampling and analytical surveys that follow the NA boundary concept are needed to establish the sustainable use and management of water resources on an international scale.

References

Cook P, Herczeg AL (2000) Environmental tracers in subsurface hydrology. Kluwer Academic Publishers, Boston, 529p

Hoefs J (2004) Stable isotope geochemistry. Springer, Berlin, 244p

Hosono T, Buapeng S, Onodera S, Yamanaka T, Shimada J, Nakano T, Taniguchi M (2009a) Sulfate and strontium isotopic variations of groundwater in the Lower Central Plain, Thailand. IAHS Publ 329:284–290

Hosono T, Robert D, Onodera S, Umezawa U, Nakano T, Taniguchi M (2009b) Cause of groundwater contamination in Jakarta alluvium volcanic fan deduced by sulfate and strontium isotope ratios. IAHS Publ 329:201–206

Hosono T, Ikawa R, Shimada J, Nakano T, Saito M, Onodera S, Lee K-K, Taniguchi M (2009c) Human impacts on groundwater flow and contamination deduced by multiple isotopes in Seoul City, South Korea. Sci Total Environ 407:3189–3197

Hosono T, Umezawa Y, Onodera S, Wang C-H, Siringan F, Buapeng S, Delinom R, Nakano T, Taniguchi M (2009d) Comparative study on water quality among Asian megacities based on major ion concentrations. In: Taniguchi M, Burnet WC, Fukushima Y, Haigh M, Umezawa Y (eds) From headwaters to the ocean: hydrological changes and watershed management. Taylor & Francis, London, pp 295–300

Hosono T, Siringan F, Yamanaka T, Umezawa Y, Onodera S, Nakano T, Taniguchi M (2010) Application of multi-isotope ratios to study the source and quality of urban groundwater in Metro Manila, Philippines. Appl Geochem 25:900–909

Hosono T, Wang C-H, Umezawa Y, Nakano T, Onodera S, Nagata T, Yoshimizu C, Tayasu I, Taniguchi M (2011a) Multiple isotope (H, O, N, S and Sr) approach elucidates complex pollution causes in the shallow groundwater of the Taipei urban area. J Hydrol 379:23–36

Hosono T, Nakano T, Shimizu Y, Onodera S, Taniguchi M (2011b) Hydrogeological constraint on nitrate and arsenic contamination in Asian metropolitan groundwater. Hydrol Processes 25:2742–2754

Hosono T, Delinom R, Nakano T, Kagabu M, Shimada J (2011c) Evolution model of $\delta^{34}S$ and $\delta^{18}O$ in dissolved sulfate in volcanic fan aquifers from recharge to coastal zone and through the Jakarta urban area, Indonesia. Sci Total Environ 409:2541–2554

Kendall C (1998) Tracing nitrogen source and cycling in catchments. In: Kendall C, McDonnell JJ (eds) Isotope tracers in catchment hydrology. Elsevier Science B.V, Amsterdam, pp 519–576

Lawrence AR, Morris BL, Gooddy DC, Calow R, Bird MJ (1997) The study of the pollution risk to deep groundwater from urban waste waters: project summary report. British Geological Survey Technical Report

Sidarto TT, Agustyanus DA, Hadiwidjojo MMP (1992) Oega Geologi Lembar Jakarta dan Kepulauan Seribu, Jawa, P3G, Bandung (in Indonesian)

Sinsakul S (2000) Late Quaternary geology of the Lower Central Plain, Thailand. J Asian Earth Sci 18:415–426

Umezawa Y, Hosono T, Onodera S, Siringan F, Buapeng S, Delinom R, Yoshimizu C, Tayasu I, Nagata T, Taniguchi M (2009) Tracing the sources of nitrate and ammonium contaminations in groundwater at developing Asian megacities, using GIS data and nitrate $\delta^{15}N$ and $\delta^{18}O$. Sci Total Environ 407:3219–3231

UNEP (2003) Groundwater and its susceptibility to degradation: a global assessment of the problem and options for management. United Nations Environmental Programme, 126p

USGS (2009) Redox conditions in selected principal aquifers of the United States. Denver Publishing Service Center, Denver

Chapter 6
A Radon-Based Mass Balance Model for Assessing Groundwater Inflows to Lakes

William C. Burnett and Natasha Dimova

Abstract Radon (Rn) in lake waters can be derived from the following sources: production by decay of dissolved Ra-226 (parent of Rn-222); diffusion from sediments; inflow from streams or rivers (especially if these surface inputs are groundwater derived); and direct groundwater inputs. Losses of radon from lakes include radioactive decay; evasion to the atmosphere; and outflows of lake water either by surface flow or recharge to underlying aquifers. We have examined the radon budget of a small lake in Florida and observed that inputs appear to be dominated by groundwater and losses via atmospheric evasion. This prompted us to develop a simple mass balance model to allow for rapid estimations of groundwater flow into lakes.

Our model applies the following assumptions: (1) the lake is well-mixed both horizontally and vertically; (2) the only radon inputs to the lake are associated with advection of groundwater; and (3) the only losses of radon are via decay and loss to the atmosphere. Preliminary studies of several "seepage" lakes in Florida have shown that they are indeed well-mixed, with no significant gradients in radon concentrations. The assumptions regarding inputs and outputs of radon also seem to hold, at least for lakes without significant surface water inflows or outflows.

In order to evaluate the groundwater-derived radon flux, we monitor the Rn concentration in lake water over time for a period long enough (usually 1–3 days) to observe changes likely caused by variations in atmospheric exchange (primarily a function of wind speed and temperature). We then attempt to reproduce the observed record by accounting for decay and atmospheric losses and by estimating the total Rn input flux using an iterative approach. Once a quasi steady-state Rn flux is evaluated by balancing the calculated outputs, we divide this flux by the measured or assumed groundwater radon concentration to determine the groundwater discharge.

W.C. Burnett (✉) • N. Dimova
Department of Earth, Ocean and Atmospheric Sciences, Florida State University, Tallahassee, FL 32306, USA
e-mail: wburnett@fsu.edu

M. Taniguchi and T. Shiraiwa (eds.), *The Dilemma of Boundaries: Toward a New Concept of Catchment*, Global Environmental Studies,
DOI 10.1007/978-4-431-54035-9_6, © Springer Japan 2012

Keywords Florida • Groundwater discharge • Lakes • Radon

6.1 Introduction

Quantifying groundwater discharge in aquatic environments can be important because of the potential significance of groundwater to the hydrologic budget, chemistry, and ecology of the receiving body of water. Groundwater discharge should occur anywhere that an aquifer with a positive head is hydraulically connected with a surface reservoir through permeable rocks or bottom sediments (Johannes 1980). Groundwater seepage can represent a significant input into the common solution lakes of central and northern Florida because of porous soils, the relatively shallow water table, and lack of surface water inlets or outlets along the ridge and upland areas (Katz et al. 1995). Because of the potential importance of groundwater to surface water quality, it is necessary to understand these interactions and to assess their role in the hydrologic budget. Unfortunately, this is typically not an easy task.

In Florida and elsewhere, it has recently become important to develop a better understanding of groundwater influences on surface water systems. To address federal requirements under the Clean Water Act (CWA), the State of Florida must identify and assess the characteristics of "impaired" surface waters that do not meet water quality standards. It is further required that Total Maximum Daily Loads (TMDLs) for these waters be established on a prioritized schedule. TMDLs are intended to set the maximum amount of a pollutant that a water body can assimilate without exceeding water quality standards. The flux of a pollutant could originate from groundwater as well as surface discharges. Within the State of Florida alone, there are thousands of such "impaired waters" as defined by the CWA. Incorporating the groundwater component of a water budget in the mass balance calculation of pollutant loading to a water body requires assessing the groundwater discharge into the impaired surface water segments. In addition, in order to achieve the water quality benefits intended by the CWA, it is critical that TMDLs be developed and implemented in a timely manner. Doing so by traditional means (e.g., detailed hydrogeologic investigations, numerical modeling) is a time-consuming and costly process. As a result, many of these assessments are being performed using sub-optimum data for possible groundwater inputs. In many cases, assessments are performed without any consideration of groundwater pathways.

In order to examine groundwater discharge into several lakes throughout a large region, a procedure must be developed that is reasonably simple as well as time and cost effective. We present here a mass balance model, based on use of naturally-occurring radon as a groundwater tracer, that we feel allows for an inexpensive yet rigorous evaluation of groundwater inputs into small seepage lakes characteristic of many parts of Florida. It is likely that this approach can be extended to small lakes in other areas as well as larger lakes in many cases.

6.2 Model Approach and Assumptions

6.2.1 *Radon as a Tracer*

The naturally-occurring radioactive gas ^{222}Rn (radon; $t_{1/2}$=3.8 d) is a good natural tracer of groundwater discharge because it is conservative and typically found at concentrations 2–3 orders of magnitude higher in groundwater than surface waters. In addition, new technology has allowed rapid and inexpensive field measurements of radon-in-water. While radon has been used to assess infiltration of surface waters into aquifers (Hoehn and von Gunten 1989; Hamada and Komae 1998), it is more powerful as a tracer of groundwater discharge into surface receiving bodies of water (Ellins et al. 1990; Cook et al. 2003; Cook et al. 2006; Mullinger et al. 2007).

Over the past several years recognition of the importance of groundwater discharge into coastal areas has increased. The radon models developed for these environments were somewhat complex because of corrections needed for tidal mixing with offshore low-radon waters as well as atmospheric evasion (Burnett and Dulaiova 2003; Burnett et al. 2006). Shallow, non-stratified lakes on the other hand present much simpler modeling situations. For lakes without any significant surface inflows or outflows, the only possible sources of radon to lakes besides the small amount contributed by dissolved ^{226}Ra (parent of ^{222}Rn) would be diffusion from sediments and groundwater discharge (either as discrete springs or disseminated seepage). Radon losses in such situations would include radioactive decay, evasion to the atmosphere, and loss by recharge into underlying aquifers.

In cases where groundwater inputs are an important part of the hydrologic cycle, groundwater discharge should dominate the radon inputs. On the other hand, radon losses in such lakes via recharge into a shallow aquifer would likely be minor as the concentration of radon seeping into sediments would be much lower than that seeping into the lake. For example if the same amount of water seeped out of one side of a lake that seeped in the other side, and the radon-in-groundwater concentration was two orders of magnitude higher than the lake water, the resulting loss of radon via recharge would only be on the order of ~1%. Thus, in these cases we have a situation where the radon budget is essentially a balance between inflows via groundwater and losses solely by decay and atmospheric evasion (Fig. 6.1).

6.2.2 *Mass Balance Model*

In a steady-state system that has a constant source of ^{222}Rn and the only loss is by decay, the inventory can be predicted by the following equation:

$$I = F \bullet \frac{1 - e^{-\lambda t}}{\lambda} \tag{6.1}$$

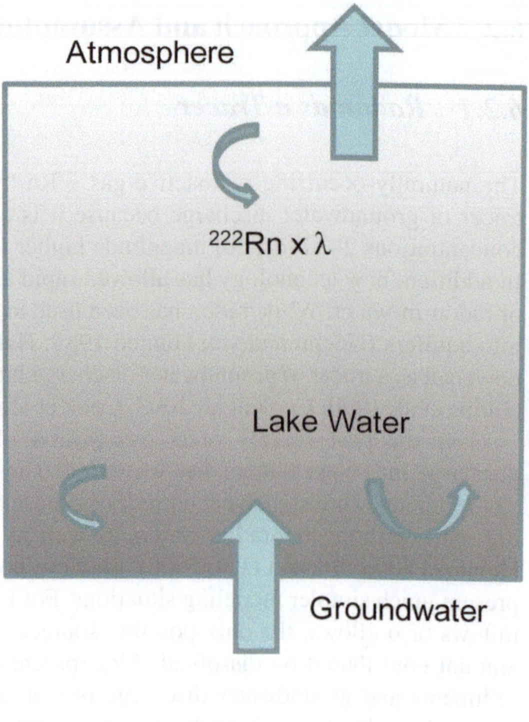

Fig. 6.1 Our model envisions radon entering lakes mainly via groundwater discharge and exiting mainly by decay and exchange with the atmosphere. We can measure how radon changes over time within lake waters and account for decay and evasion. We can thus estimate radon (and groundwater) inputs by a simple flux-by-difference approach

where I = inventory of ^{222}Rn (dpm/m^2; dpm = disintegrations per minute); F = ^{222}Rn flux (dpm/m^2 h), and λ = ^{222}Rn decay constant (7.56×10^{-3} h^{-1}). For a shallow lake that is not stratified, the ^{222}Rn inventory would simply be the concentration (dpm/m^3) times the water depth (m). After several half-lives of ^{222}Rn, (1) becomes

$$I = F \bullet \frac{1}{\lambda} \tag{6.2}$$

where $1/\lambda$ = the mean life of ^{222}Rn (~5.5 days). Thus, if we can assume that the flux of radon (groundwater) is reasonably constant over a period on the scale of days to weeks we can predict what the inventory (or concentration by dividing the inventory by the water depth) should be without any losses other than by decay. However, it is always the case that radon will be lost to the atmosphere so this must be included in the balance. Fortunately, a considerable amount of effort has gone into developing models to predict gas exchange between water and air (e.g., see review in MacIntyre et al. 1995). The general equation that relates the flux of a gas (F_{atm}) across a water–air interface is:

$$F_{atm} = k * (C_w - \alpha * C_a) \tag{6.3}$$

where k = the gas transfer coefficient or piston velocity (m/day), C_w and C_a = the concentration (dpm/m^3) of the gas in water and air respectively, and α = Ostwald's solubility coefficient, which is temperature dependent. The concentrations of radon in both the air and water phases can be measured relatively easily and the solubility can be determined from a temperature-dependent relationship. The gas transfer coefficient, k, is the key to determining reasonable fluxes. This parameter is a function of the physical processes at the air–water interface, especially the turbulence and kinematic viscosity (ν) of the water, and the molecular diffusion coefficient (D_m) of the gas (1.16×10^{-5} cm²/s at 20°C for radon). Based on a number of studies, empirical equations which relate k to temperature and to wind speed as a measure of turbulence have been proposed. We have found through several years of experience that the following empirical equation presented by MacIntyre et al. (1995) appears to work very well:

$$k(600) = 0.45u_{10}^{1.6}(Sc/600)^{-0.5} \tag{6.4}$$

where u_{10} = the wind speed at 10 m height above the water surface and Sc is the Schmidt number (the ratio of the kinematic viscosity to the molecular diffusion coefficient; i.e., $Sc = \nu/D_m$). Both ν and D_m can be adjusted for temperature changes by well-known relationships. Sc is divided by 600 to normalize k to CO_2 at 20°C in freshwater.

On a few occasions we have been able to estimate radon losses from the sea surface experimentally, via comparison to ^{224}Ra trends, an isotope of radium that has a very similar half-life to radon and often has similar sources, yet no gas phase. These comparisons, made in Thailand (Dulaiova and Burnett 2006) and Brazil (Burnett et al. 2008), showed good agreement to the theoretical treatment.

We are thus in a position where we can estimate groundwater discharge into lakes by first measuring radon concentration trends over time within a lake and then modeling those trends. We will assume here that a lake is well-mixed both horizontally and vertically so that we can measure the temporal trends at any point in the lake. We further assume that the only significant source of radon is via groundwater transport and that the only significant loss besides decay is atmospheric evasion. We can thus model the trends in the radon concentration in the lake by: (1) measuring or making reasonable estimates of the controlling parameters (Table 6.1); and (2) adjusting the radon flux (via groundwater) to balance the observed trends. Once we have a good estimate of the steady-state flux, we can convert this to a water flux by simply dividing by the measured radon-in-groundwater concentration. We consider this flux to be "quasi steady-state" as it is reasonable to expect relatively constant fluxes over time scales of days to weeks when environmental conditions are not punctuated by storms or other unusual conditions. Seasonal variations reflecting dry and wet conditions would clearly result in longer-term variations.

Table 6.1 Parameters used to evaluate the theoretical budget of ^{222}Rn in shallow lakes as a balance between groundwater inputs with decay and atmospheric losses

Parameter	Units	Evaluation
Atmospheric ^{222}Rn	dpm/m^3	Measure or assume value; model very insensitive to Rn-in-air variations
Wind speed (u_{10})	m/s	Measure or download from nearby weather station
Water/air temperatures	°C	Continuous measurement with probes, data loggers, weather station
Molecular diffusion (Dm) of Rn in water	cm^2/s	1.16×10^{-5} cm^2/s, adjust for temperature
Viscosity (ν)	cm^2/s	Calculate based on water density, temperature
Schmidt number (Sc)	Dimensionless	Calculate via ν/D_m
Water depth	m	Measure
Decay constant ^{222}Rn (λ)	7.56×10^{-3} h^{-1}	Constant
Steady-state flux (F_{Rn})	dpm/m^2 h	Adjust to balance losses, match observations

6.3 Site Description and Experimental Approach

The results presented in this chapter are from Lake Haines, a small lake in Polk County, central Florida (Fig. 6.2). The lake has a surface area of approximately 1.23×10^6 m^2 and an average depth of 1.8 m. The land use in the area is fairly high-density suburban with significant development of retirement and golf-oriented living communities. Other studies have shown that the lake is experiencing progressive eutrophication, most likely from nutrient loading, and is considered an "impaired water" (Whitmore and Brenner 2002).

In order to characterize the radon trends in the lake water, we ran continuous measurements over periods long enough (generally a few days) that we would expect to see variations due to changing environmental conditions (temperature, wind speed) that would affect radon losses. The radon instrumentation (Durridge Inc. RAD-7) was setup as a mooring on private docks on the lake. The commercial radon-in-air instrumentation has been modified to measure radon-in-water automatically and continuously (Burnett et al. 2001). Lake water (from ~30 cm depth) was delivered using a submersible pump into the top of an air–water mixing chamber (commercialized as a "RAD-AQUA" system), where the radon within the air loop that circulates through the RAD-7 equilibrates with the radon in the water stream. By measuring the radon concentration in the air loop together with the temperature in the mixing chamber (via a HOBO temperature probe), one can calculate the amount of radon within the water based on its solubility. We used integration times of 30 min and were able to achieve results with an uncertainty in the radon concentration of about ±10% (1σ). This automated system was used in conjunction with continuous temperature/conductivity logging systems (Van Essen Instruments Data Divers) as well as a HOBO Water Level Meter (Onset Corp.). While the time-series experiment was running, shallow groundwater and pore water samples were collected in and around the lake from a monitoring well and via a push-point

Fig. 6.2 Aerial view of Lake Haines, Florida showing the locations of three moorings set up to monitor radon and other water quality parameters

piezometer system to estimate the radon content of the groundwater. These samples were processed for ^{222}Rn concentrations via a modified "RAD-H2O" technique (Lee and Kim 2006). A single 20-liter sample of lake water was collected and concentrated onto Mn-impregnated acrylic fiber in the field and later measured for ^{226}Ra in the laboratory by a radon emanation technique (Peterson et al. 2009). Sediment samples were also collected at a few sites to characterize the groundwater "end-member," i.e., the characteristic ^{222}Rn concentration in the groundwater discharging to the lake. Sediment samples were analyzed for ^{226}Ra by gamma spectrometry and sediments were reacted with water in closed vessels in the laboratory to evaluate the equilibrium concentration of radon following the techniques described in Corbett et al. (1998).

6.4 Results and Discussion

6.4.1 Mooring Results: March 2008

We established a fixed point mooring on a dock at a private golf resort on the east shore of Lake Haines and monitored radon, temperature, and conductivity from 17:30 March 24 to 11:00 March 28 over time (30-min interval for radon). Radon in air was also monitored with a separate RAD7 unit setup at the golf club. Air temperatures and wind speeds were downloaded from a nearby weather station via the internet portal Weather Underground (http://www.wunderground.com). This experiment was performed at a time of year in Florida when there can be dramatic shifts in the air temperature as cold fronts still move in from the north while solar warming occurs during the longer Spring days. This turned out to be the case during this period with observed temperatures in our mixing chamber ranging from a low of 7.2°C to a high of 31.1°C. The mixing chamber is where the radon-in-water equilibrates with the radon-in-air, so the temperature in the chamber is the temperature applied when calculating radon concentrations. The extreme temperature range during this mooring was a benefit to our objective as it provided a robust radon-over-time record (Fig. 6.3) showing distinct diurnal variations with radon higher in the evening and early morning hours (low temperatures and wind speeds) and lower concentrations during the day time hours (higher temperatures and wind speeds). There was also a general downward trend in the radon concentrations over the several days of the experiment. This was likely related to a general warming of the lake waters during this period making the radon less soluble.

We modeled the data by first initializing the run with a starting concentration or inventory (since we consider the water depth in the lake to be constant over the period of our measurements, one can work in either inventories or concentrations). We used the first valid measurement point as a starting point for the model. Presumably, if we ran the model long enough (a few half-lives of radon), the starting concentration selected would not matter as the system would have time to adjust to any starting value.

We then corrected the value from the previous time step for decay, a very small correction because of the short 30-min interval between measurements. We assumed a radon flux and calculated the inventory gained during each measurement interval via (6.1) (the assumed flux was later adjusted to better fit the observed data). The rate of loss of radon via atmospheric evasion was calculated for each interval according to (6.3) and (6.4) with relevant adjustments made for viscosity (ν) and molecular diffusion (D_m) depending upon temperature. The piston velocity is mostly a function of the wind speed and the Sc parameter, which ultimately is a function of temperature. The atmospheric flux is converted to an equivalent inventory lost in the same manner as the input flux is converted to inventory (6.1). The net inventory for each time step is thus the result of combining the value from the previous time step with the sum of the computed input (assumed via groundwater) minus the outflow by atmospheric evasion (radioactive decay is already accounted for). The resulting modeled inventory

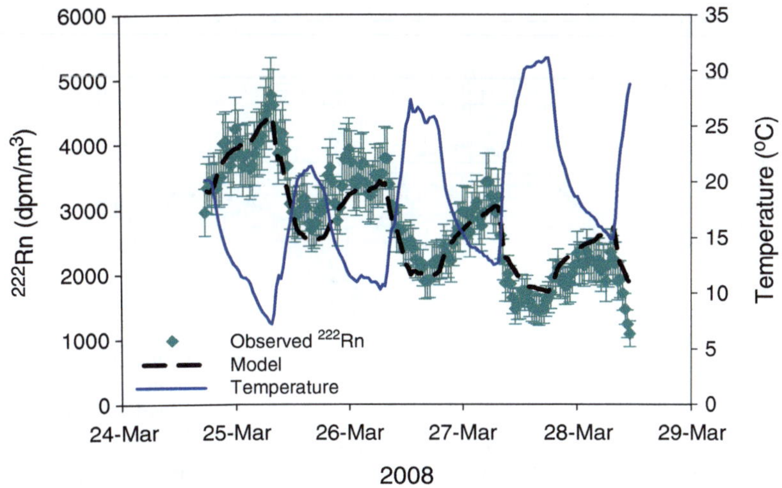

Fig. 6.3 Plot of observed and modeled radon concentrations as well as temperature variations from the dock at the Golf Club on the east shore of Lake Haines, Florida. The observed values were monitored from March 24 to 28, 2008

over time is based on an assumed flux, which can be adjusted to the quasi steady-state value needed to best match the observations. All terms in the model are observations of measured parameters or calculations based on first principles.

In the case of the March 2008 experiment, we found an excellent fit (based on the lowest rms value between the observed and modeled values) using an input flux of 44 dpm/m^2 h (equivalent to a steady-state ^{222}Rn concentration of 4500 dpm/m^3). Using a radon-in-groundwater value of 211,000 dpm/m^3, measured from a monitoring well adjacent to the lake, this is equivalent to an average seepage rate of 0.50 cm/day (0.50 cm^3/cm^2 day).

6.4.2 *Mooring Results: December 2008*

In order to evaluate our assumption of a well-mixed lake, we returned to Lake Haines in December 2008 and established 3 moorings at 3 different locations: (1) the same golf course location tested in March; (2) a boat ramp on the opposite (west) side of the lake; and (3) a dock at a private house in the southwest corner of the lake (locations shown in Fig. 6.2). The observed results from the three locations appear to be very similar (Fig. 6.4). The match between the golf course and the boat ramp locations is particularly good with overlapping data throughout the over 1.5 day period. We could not get the mooring started at the private house until the second day so the record at that location is shorter. The private house results were also consistently somewhat higher in radon concentration than the other two

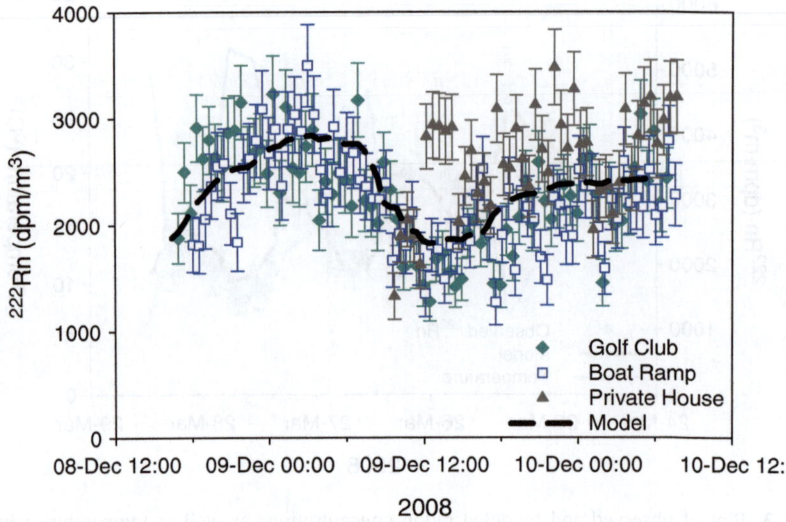

Fig. 6.4 Plot of observed radon concentrations from three sites at Lake Haines: the dock at the Golf Club; a boat ramp on the western side; and a dock at a private house in the southwest corner of the lake. The model result for the Golf Club site is shown

stations, perhaps because the dock there was part of a boat house and thus in a more sheltered environment.

Analyzing the two records that spanned the entire interval produced quasi steady-state radon flux estimates of 48 and 51 dpm/m^2 h for the golf course and boat ramp locations respectively. These estimates are close to each other and close to that estimated from the March record at the golf course (44 dpm/m^2 h). Again using an estimated groundwater concentration of 211,000 dpm/m^3, average seepage rates of 0.55 and 0.57 cm/day are derived for the golf course and boat ramp sites, respectively.

6.4.3 Summary of Results

When scaled to the area of the lake, an average seepage rate of 0.50 cm/day (March 2008 estimate) is equivalent to 6200 m^3/day groundwater inflow. Since we are assuming that the lake is well-mixed, the estimated seepage rate should represent an inflow over the entire area of the lake bottom. This may not actually be the situation as preferential flows through sandier areas are likely. As long as the lake is well-mixed, and our results indicate that it is, the exact sites of inflow do not matter. At the estimated rate of groundwater discharge, about 0.3% of the lake's estimated total volume (~2.2×10^6 m^3) is replaced each day. The residence time of the lake water if groundwater is the only input would thus be approximately 1 year. Independent data obtained by a consulting company (PBSJ 2009) that performed seepage meter measurements at several locations in Lake Haines and at various times during the same year (monthly from June to November 2008) indicated an average seepage rate of 0.65 cm/day, slightly higher than our 0.50–0.57 cm/day estimates. While seepage meters are useful as the only means available for direct measurement of groundwater discharge, the devices only cover a small (~0.5 m^2) area of lake bed and thus many measurements are needed to obtain representative results. The radon approach, on the other hand, integrates the signal over a very wide area (the entire lake if our well-mixed assumption is correct) and thus is an excellent way to deal with the spatial variability that is usually associated with groundwater discharge.

In summary, our radon tracing approach for small seepage lakes consists of deployment of a mooring to measure radon concentration and temperature in the lake water over time. Parallel measurements of wind speed, radon-in-air, and groundwater radon concentrations then allow for modeling observed variations in radon inventories within the lake based on a mass balance of radon inputs via groundwater and losses by decay and evasion to the atmosphere. Our results from Lake Haines (Florida) seem reasonable and are in close accord to more laborious seepage meter measurements. It is likely that the same general approach can be used in larger lakes, especially those that do not have significant surface water inputs.

Acknowledgments The authors thank Bochao Xu and Natalia Tobon for assistance with the fieldwork. Personnel from the Florida Department of Environmental Protection (FDEP), especially Richard Hicks and Xueqing Gao, were instrumental in initiating this research. We also acknowledge FDEP for the financial support of this research through contract #GW270.

References

Burnett WC, Dulaiova H (2003) Estimating the dynamics of groundwater input into the coastal zone via continuous radon-222 measurements. J Environ Radioact 69:21–35

Burnett WC, Kim G, Lane-Smith D (2001) A continuous radon monitor for assessment of radon in coastal ocean waters. J Radioanal Nucl Chem 249:167–172

Burnett WC et al (2006) Quantifying submarine groundwater discharge in the coastal zone via multiple methods. Sci Total Environ 367:498–543

Burnett WC, Peterson R, Moore WS, de Oliveira J (2008) Radon and radium isotopes as tracers of submarine groundwater discharge—results from the Ubatuba, Brazil SGD assessment intercomparison. Estuar Coastal Shelf Sci 76:501–511

Cook PG, Favreau G, Dighton JC, Tickell S (2003) Determining natural groundwater influx to a tropical river using radon, chlorofluorocarbons and ionic environmental tracers. J Hydrol 277:74–88

Cook PG, Lamontagne S, Berhane D, Clark JF (2006) Quantifying groundwater discharge to Cockburn River, southeastern Australia, using dissolved gas tracers ^{222}Rn and SF_6. Water Resour Res 42:1–12

Corbett DR, Burnett WC, Cable PH, Clark SB (1998) A multiple approach to the determination of radon fluxes from sediments. J Radioanal Nucl Chem 236:247–252

Dulaiova H, Burnett WC (2006) Radon loss across the water–air interface estimated from ^{222}Rn-^{224}Ra. Geophys Res Lett 33:L05606. doi:10.1029/2005GL025023

Ellins KK, Roman-Mas A, Lee R (1990) Using ^{222}Rn to examine groundwater/surface discharge interaction in the Rio Grande de Manati, Puerto Rico. J Hydrol 115:319–341

Hamada H, Komae T (1998) Analysis of recharge by paddy field irrigation using ^{222}Rn concentration in groundwater as an indicator. J Hydrol 205:92–100

Hoehn E, von Gunten HR (1989) Radon in groundwater: a tool to assess infiltration from surface waters to aquifers. Water Resour Res 25(8):1795–1803

Johannes RE (1980) The ecological significance of the submarine discharge of groundwater. Mar Ecol Prog Ser 3:365–373

Katz BG, Plummer LN, Busenberg E, Revesz KM, Jones BF, Lee TM (1995) Chemical evolution of groundwater near a sinkhole lake, northern Florida-2. Water Resour Res 31:1565–1584

Lee JM, Kim G (2006) A simple and rapid method for analyzing radon in coastal and ground waters using a radon-in-air monitor. J Environ Radioact 89:219–228

MacIntyre S, Wanninkhof SR, Chanton JP (1995) Trace gas exchange across the air–sea interface in freshwater and coastal marine environments. In: Matson PA, Harris RC (eds) Biogenic trace gases: measuring emissions from soil and water. Blackwell Science, London, pp 52–97

Mullinger NJ, Binley AM, Pates JM, Crook NP (2007) Radon in chalk streams: Spatial and temporal variation of groundwater sources in the Pang and Lambourn catchments, UK. J Hydrol 339:172–182

PBSJ (2009) Winterhaven chain of lakes groundwater seepage study. Consultant's Report to FDEP, January 2009

Peterson RN, Burnett WC, Dimova N, Santos IR (2009) A comparison of measurement methods for radium-226 on manganese-fiber. Limnol Oceanogr Meth 7:196–205

Whitmore TJ, Brenner M (2002) Paleolimnological characterization of pre-disturbance water quality conditions in EPA/DEP-defined Florida lake regions. Final Report Florida Department of Environmental Protection 77 p

Part III
Transboundary Linkage of Land and Ocean

Chapter 7
Introduction

Takeshi Nakatsuka

Keyword Transboundary material flows

From the academic viewpoint, we can divide the biosphere—the thin layer supporting life on Earth—into the atmosphere, land and oceans. For each of these, specific fields of science have been established over the last several centuries, with great advances made during recent decades. In the realm of atmosphere, meteorology has been established. In the realm the oceans, oceanography and fisheries were created. On the topic of land, many academic disciplines such as hydrology, forestry, agriculture have been established. While each science has been developing foundational knowledge in its specific realm for a long time, the rise of global environmental issues inevitably demands the rise of trans-boundary studies which bridge atmosphere, ocean and land.

The linkages between atmosphere and the other two realms have been investigated relatively well. This is likely due to the fact that both land and ocean are exposed to the atmosphere at huge areas around the planet, and physical–chemical interactions occurring at the atmosphere-land and atmosphere-ocean interfaces can be recognized quite easily. On the other hand, land–ocean linkages, especially continental and open ocean linkages, have not been examined frequently. Because coastal zones are very productive and important for fisheries, they have been studied intensively. However, the role of coastal zones as connections between land and ocean has not been understood in depth.

In fact, many important materials are transported through coastal zones from land to ocean, and sometimes vice versa. The material flows do not only support ecosystems in coastal zones but also influence inland and open ocean ecosystems.

T. Nakatsuka (✉)
Graduate School of Environmental Studies, Nagoya University,
Nagoya 464-8601, Aichi, Japan
e-mail: nakatsuka.takeshi@f.mbox.nagoya-u.ac.jp

M. Taniguchi and T. Shiraiwa (eds.), *The Dilemma of Boundaries: Toward a New Concept of Catchment*, Global Environmental Studies,
DOI 10.1007/978-4-431-54035-9_7,

Two kinds of land–ocean linkages were presented with respect to different types of material flow (iron and fish). While there are other important key materials connecting land and ocean, examinations of the two materials have clearly illustrated the importance of trans-boundary and trans-disciplinary studies necessary for understanding and protecting the unexpected linkages between land and ocean.

Chapter 8 proposed the new concept of the “Giant Fish Breeding Forest” as an enlarged version of the traditional Japanese concept of the fish breeding forest (Uo-Tsuki-Rin). In this new concept, dissolved iron plays the key role. Although iron is one of the indispensable elements for growth of marine phytoplankton, iron is minimally dissolved in seawater and iron supplied from rivers precipitates rapidly in coastal zones. Therefore, it has been widely accepted that river iron can never reach the open ocean and that oceanic phytoplankton depends only on the atmospheric iron transported as aerosol particles from arid land surface. Chapter 8 showed a novel iron transport route involving river and ocean currents connecting the Amur River watershed in northeast Asia and the Oyashio region in the northwest Pacific through the Sea of Okhotsk. The reason why Amur River iron can be transported far into open ocean areas is that there is an intermediate water ventilation system in the Sea of Okhotsk supported by seasonal sea ice and dense brine water formation. Because a large amount of seasonal sea ice is formed on the continental shelf around the Amur River mouth every winter, the precipitated iron can be entrained in the dense water rejected from sea ice penetrating into the intermediate layer of 200–500 m. Due to the scarce biological activity in the intermediate layer, some portion of the iron can survive without being utilized or scavenged and can reach the Pacific Ocean through the Kuril straits where the iron can move into the surface layer of Oyashio water on the east of the Kuril archipelago through strong tidal mixing in the straits. Because the Oyashio region is known as one of the most productive fisheries in the world, this novel iron transport system induces great implications for our society. Therefore, Chap. 8 proposed the new concept of the Amur River watershed as a “Giant fish breeding forest” for Oyashio.

While the forest plays a surprisingly important role in iron transport, the most important component in the Amur River watershed for dissolved iron discharge is not the forest but the wetland. Chapter 9 demonstrated that wetlands around the middle and lower reaches of the Amur River can provide the Amur River with dissolved iron sufficient to explain the very high dissolved iron concentration in Amur River water using an integrated ground and surface water hydrological and biogeochemical model covering all of the Amur River watershed. Meanwhile, Chap. 9 proved, through a regional high-resolution ocean circulation model, that intermediate water in the Sea of Okhotsk can actually transport iron from the Amur River mouth to the Pacific Ocean within the rapid time.

Chapter 10 showed a linkage in the opposite direction between land and ocean involving anadromous fish such as salmon. Although river water flows from land to ocean discharging huge amounts of terrestrial materials to the coastal zone, anadromous fish can transport marine-derived materials, especially nutrients such as nitrogen and phosphorous, upward onto the land. Because nutrients are a limiting factor for forest growth, this upward supply of marine-derived nutrients (MDN) may play a significant role for ecosystems around rivers. Quantitative estimates of MDN were first conducted in North America, mainly focusing on Pacific salmon, but there have been few studies in the Russian Far East. Based on the data of escapement of adult Pacific salmon to spawning areas in the Russian Far East, Chap. 10 showed preliminary estimates of how much MDN are annually uploaded onto terrestrial ecosystems from the Northern Pacific.

Although iron and fish must play significant roles in marine and terrestrial ecosystems, all of these material transport systems are now facing threats owing to human activities. The common factors damaging these material transport systems are dam construction and modification of river channels. Dams are often important for local people as the key components of water management systems and/or electric power supply. However, dams not only prevent the anadromous fishes from passing through them, as indicated by Chap. 10, but they also weaken iron discharge from wetlands by stabilizing river water levels and drying flood plains. In the case of the "Giant fish breeding forest" proposed in Chap. 8, there are other problems as well. Chapter 9 demonstrated that land use change from wetlands to farmland results in a significant decrease of iron discharge from the Amur River. Global warming is also reducing the sea-ice driven ventilation of intermediate water and the amount of iron transport as calculated in Chap. 9.

Protecting the material transport systems between land and ocean is very challenging because it requires linking many stakeholders. Material transport between land and ocean is often unidirectional and the asymmetric relationship between upstream and downstream stakeholders usually prevents concerned stakeholders from coming to agreement. In the case of the "Giant fish breeding forest," multiple countries, including China, Russia, Mongolia and Japan, are located in this system asymmetrically. This situation presents an extremely difficult environmental management issue.

Japanese Uotsukirin (fish breeding forest) concept as described in Chap. 13 may provide good guide for future collaboration among asymmetrically arrayed stakeholders even beyond national boundaries. All over Japan, fishermen are now planting trees in the forests located upstream of their fisheries, collaborating with foresters and farmers for the overall improvement of their livelihoods and environment, including land–ocean linkage systems. We hope that these kinds of collaborations among different stakeholders develop not only within regional or national scales but also in international scales through some trans-disciplinary integration of environment, economy and politics.

Chapter 8
"Giant Fish-Breeding Forest": A New Environmental System Linking Continental Watershed with Open Water

Takayuki Shiraiwa

Abstract The Amur-Okhotsk Project (AOP) introduced a new global environmental concept referred to as the "giant fish-breeding forest" (GFBF) by expanding the traditional Japanese idea of uotsuki-rin (fish-breeding forest), which related upstream forests with the coastal ecosystem both physically and conceptually. The AOP found that primary production in the Sea of Okhotsk and the Oyashio region depended on dissolved iron transported from the Amur River and its watershed. Therefore, the Amur River basin can be recognized as the GFBF of the Sea of Okhotsk and the Oyashio region. This hypothesis presents new perspectives on global environmental issues: an ecological linkage between the continent and the open sea, relationships between stakeholders who are not necessarily depended each other in the system, and the identification of environmental common ground across coast lines and complex international boundaries. Multidisciplinary approaches are indispensable in studying and conserving the GFBF because stakeholders need to understand how to achieve a sustainable marine ecosystem in the Sea of Okhotsk and Oyashio region without limiting human activity on land. Connecting less dependent stakeholders could be a first step in coping with complicated environmental issues. We attempt to visualize socio-economic relationships inside the GFBF system to demonstrate how stakeholders are related to each other unconsciously. Establishment of the concept will help bring together people who have been separated for many years under political tensions.

Keywords Amur River • Dissolved iron • Fish-breeding forest • Open sea • Sea of Okhotsk • Uotsuki-rin

---

T. Shiraiwa (✉)
Institute of Low Temperature Science, Hokkaido University,
Kita-19, Nishi-8, Kita-ku, Sapporo 060-0819, Japan
e-mail: shiraiwa@lowtem.hokudai.ac.jp

M. Taniguchi and T. Shiraiwa (eds.), *The Dilemma of Boundaries: Toward a New Concept of Catchment*, Global Environmental Studies,
DOI 10.1007/978-4-431-54035-9_8,

8.1 Introduction

Those employed in fisheries in coastal and estuarial areas of Japan have historically protected forests in the drainage basins bordering their fishery fields. The forest was called uotsuki-rin (the fish-breeding forest) and the existence of such forest was believed to improve the condition for fish growth in the coastal areas by providing various nutrients to the coast. In the late 1970s, professional fishermen noticed that coastal zones were seriously damaged and the areas were no longer productive. Pioneering fishermen believed that such ecological deterioration in coastal and estuarial areas was caused by rapid changes in the land surface of the adjacent drainage basins through excessive land-development starting during the Japanese economic growth in the 1960s. In Hokkaido and Miyagi prefectures, leaders of fishery groups advocated reforestation in the adjacent drainage basins to restore a rich ecological environment in coastal and estuarial areas (e.g., Hatakeyama 1994; Yaginuma 1999). The movement came to be known as uotsuki-rin undoh (the fish-breeding forest movement).

Practically, it was impossible to know the extent to which the material discharged from rivers affected the marine ecosystem, the amount and kinds of materials which were exported under various conditions of land surfaces, and the distances to which terrestrial materials affect the off-shore marine ecosystem. Matsunaga et al. (1984, 1998) conducted one of the first studies to indicate a relationship between riverine organic-iron complex and phytoplankton in an estuarial area. Since then, an increasing number of studies have shown that the land–ocean linkage of material transport through rivers plays a key role in the growth of fish populations in estuarial and coastal areas (e.g., Yamashita and Tanaka 2008). The mechanisms behind the linkage are not yet fully understood; thereby, several international and domestic research organizations have begun tackling these issues with a multidisciplinary approach (e.g., Connected Rings of Forest–Human Habitation–Marine by the Field Science Education and Research Centre of Kyoto University; Land–Ocean Interactions in the Coastal Zone by the International Geosphere Biosphere Program/International Human Dimensions Program).

In contrast to the relationship between the land-surface and the adjacent estuarial and coastal areas, it is widely accepted that the relationship between the land-surface and open waters is minimal. This is because riverine materials flocculate in estuarial areas and then accumulate on the continental shelves without moving to the open waters. Our project reconsiders on this conventional idea.

The northern North Pacific is known to be high in nutrients and low in chlorophyll (HNLC); dissolved macro nutrients (nitrate, phosphate and silicate) in the surface water cannot fully be utilized by phytoplankton because of the low availability of iron. Iron is usually supplied to estuarial and coastal regions from the land surface and it is difficult for iron to be transported to the remote central area of the northern North Pacific. This idea was proposed by Martin et al. (1989) and is called the iron limitation hypothesis. The idea is now being confirmed by a mesoscale in situ iron enrichment experiment in the North Pacific (Tsuda et al. 2003).

The neighboring Sea of Okhotsk is also characterized by sufficient nutrients supplied by the winter convective mixing of surface and deep waters. The Sea of Okhotsk is, however, not a HNLC region. This is probably because sufficient dissolved iron is transported from the Amur River. The Amur River, including major tributaries like the Shilka, Argun, Zeya, Bureya, Songhua Jiang (Sungari) and Ussuri Rivers, is 4,444 km long and has a drainage area of 2,129,700 km^2 (Simonov and Dahmer 2008). The major part of the drainage area is covered by boreal forest, mixed forest and swamps. The lower part of the drainage area is cultivated land and major cities such as Blagoveshchensk, Harbin, Khabarovsk and Komsomolsk-na-Amure. The relatively less developed Amur River basin enables the river to transport various kinds of terrestrial materials to the Sea of Okhotsk. Of particular importance is dissolved iron, which is considered to originate in an anoxic environments such as swamps.

The Sea of Okhotsk was studied intensively from 1997 to 2002 during the Core Research for Evolutional Science and Technology project led by Prof. Masaaki Wakatsuchi of the Institute of Low Temperature Science, Hokkaido University. It was found that two oceanographic mechanisms transport the water and materials from the mouth of the Amur River to the northern North Pacific. One is the East Sakhalin Current, which is a western boundary current along the east Sakhalin coast. A part of this current flows eastward as far as Bussol Strait to enter the northern North Pacific. The volume transport of this current is estimated to be in the range 3–10 Sv with an increase from summer to winter (Ohshima et al. 2002; Mizuta et al. 2003). The other mechanism transports water and material from the coastal area near the Amur River to the Sea of Okhotsk and further to the northern North Pacific including the Oyashio region (Nakatsuka et al. 2002, 2004). At the bottom of the northwestern continental shelf of the Sea of Okhotsk, very cold dense water due to brine rejection forms in winter. This water is characterized by enormously high turbidity due to tidal mixing and is transported to the intermediate depth by the East Sakhalin Current. By these two mechanisms, the water and material of the Amur River are effectively transported to the southern part of the Sea of Okhotsk and further to the northern North Pacific. Although no measurement of iron concentrations were conducted during Wakatsuchi's project, it is highly probable that the riverine iron can be transported by the above mechanism to the northern North Pacific where iron is the key element controlling phytoplankton growth (Fig. 8.1).

8.2 Aims of the Amur-Okhotsk Project (AOP)

If the above-mentioned mechanism is the case, the Amur River basin plays a crucial role in determining the biomass production both in the Sea of Okhotsk and the northern North Pacific including the Oyashio region. This relation reminds us of the Japanese concept of uotsuki-rin (the fish-breeding forest: see Chapter 13). However, the Amur-Okhotsk-Oyashio linkage is much stronger than that in the conventional concept. More importantly, this is the first attempt to relate the continental-scale

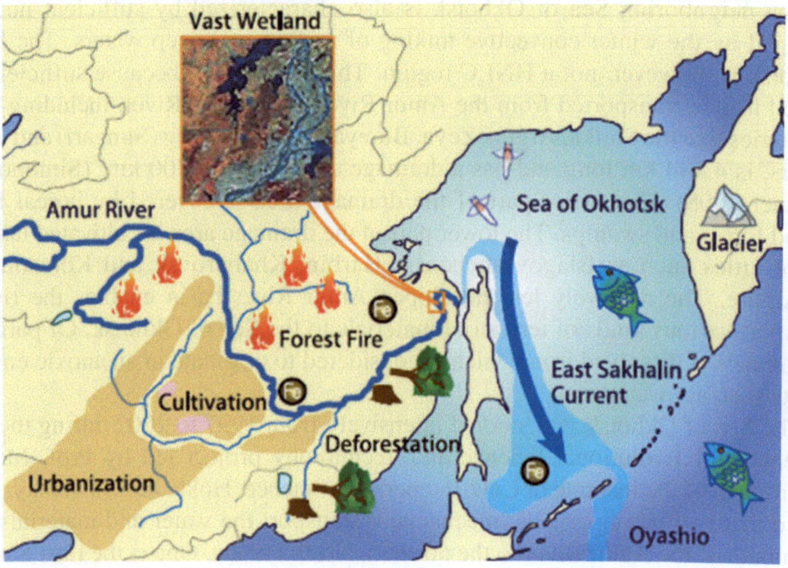

Fig. 8.1 Concept of the Amur-Okhotsk Project

terrestrial environment with open waters. Therefore, we refer to the idea as kyodai uotsuki-rin kasetsu (the "giant fish-breeding forest" (GFBF) hypothesis) and the verification of the hypothesis constitutes the first part of the AOP.

The dissolved iron mainly forms as a complex of iron and fulvic acids originating from forests and swamps in the basin (Matsunaga et al. 1998). The processes of the formation of the dissolved iron, its transportation to the river, and its delivery to the ocean are still open to question. It is, nevertheless, clear that changes in the land surface and river discharge affect the flux of the dissolved iron significantly, because the land surface and river constitute the source and method of dissolved iron inputs. The change in the dissolved iron flux may affect biomass production in the Sea of Okhotsk and the adjacent Oyashio region in the long run. Verifying the impact of terrestrial anthropogenic disturbances in the Amur River basin on primary production in the Sea of Okhotsk and Oyashio region is the second part of this project.

The Amur River drainage was developed after the end of the nineteenth century in Russia (Ganzey 2005). In China, i.e. the Songhua Jiang River basin, intensive human activity dates back several hundred years. An accelerated human impact became more obvious from the middle of the twentieth century on both sides of the Amur River. The area is disturbed currently by various anthropogenic and natural impacts such as forest fires, deforestation, agricultural and industrial activities, flooding and drought. Land-use changes in the Amur River drainage area, therefore, might have caused or may cause significant changes in the flux of dissolved iron, which might have or may result in biomass production changes in the ocean.

The Sea of Okhotsk and the northern North Pacific are known to be among the most productive oceanic areas in the world. Approximately half of the domestic sea product of Japan is from this area. Therefore, the ecosystem and environment of the Sea of Okhotsk and the adjacent Oyashio region are important not only with respect to the environment but also the economy of Japan.

Moreover, recent climatological analysis showed that sea–air CO_2 exchange in Oyashio and its adjacent areas was a unique node where intensive exchange in pCO_2 occurred (Takahashi et al. 2002). This indicates phytoplankton growth in this region is important not only to the sea product but also the global climate.

Therefore, it is important to clarify a robust relationship between conditions of the Amur River basin and ecosystems in the Sea of Okhotsk and Oyashio region to predict land-use impacts on the future marine primary productivity in the ocean. This information will be helpful not only for people concerned with uotsuki-rin, but also for people consuming sea products. In this context, the Amur River basin and the Sea of Okhotsk and Oyashio region are the most typical and extensive examples in the world.

As mentioned previously, the conservation of uotsuki-rin was begun by fishermen who profited from this system. The idea was then accepted nationwide and there began regional-scale attempts to protect uotsuki-rin by various stakeholders such as fishermen, citizens and municipal officers. The third part of this project therefore presents an idea of how we can conserve the GFBF, which encompasses multiple countries, exclusive economic zones and open waters.

In addition to physical boundaries, the Amur River has been the site of political boundaries between China and Russia since the "Treaty of Aigun" and "Convention of Peking" signed by the two countries in 1858 and 1860, respectively. Since the two agreements were considered to be unequal treaties, the boundary was rather unstable until the two countries finally agreed to define the boundary in 2004. This history has made the Amur River one of the most difficult rivers to monitor for conservation purposes. In practice, there was no formal joint-monitoring program between the two countries until a notorious accident involving a petrochemical company in the Chinese province of Jiling in 2005. This accident significantly polluted the Songhua River, the largest tributary of the Amur River (UNEP 2006).

The Sea of Okhotsk has been a political hot spot between Russia and Japan. Owing to territorial conflict, it was practically impossible for Japan to monitor environmental problems in the Sea of Okhotsk during the last half of the twentieth century. A collaborative effort between Russia and Japan for the environmental conservation of the Sea of Okhotsk is urgently needed owing to increasing activities related to oil mining and natural gas exploration in the Sea of Okhotsk and its vicinity. Clearly, the GFBF is of local as well as worldwide importance and the elaborabation of the concept will contribute to the solving of practical problems several countries are facing.

8.3 Results of the AOP

Major achievements of the Amur Okhotsk Project (2005–2009) are described by answering the five essential questions of the project.

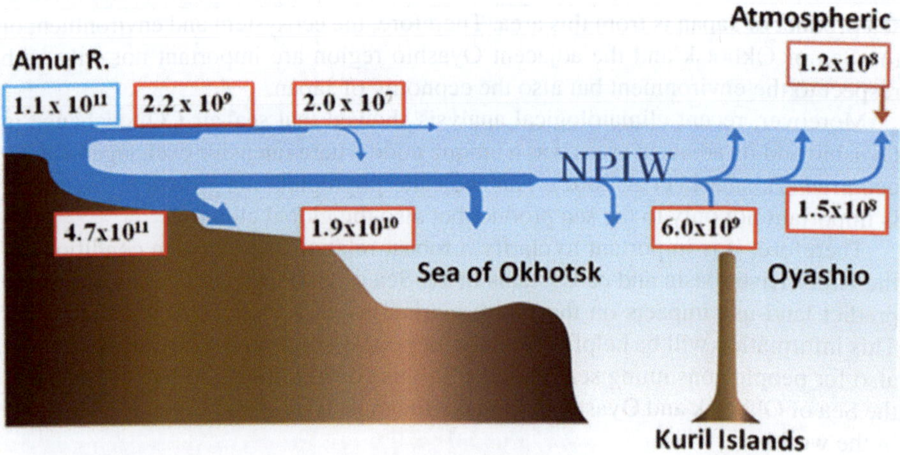

Fig. 8.2 Quantitative evaluation of Fe transport efficiency (g/year) in the Sea of Okhotsk (modified from Nakatsuka et al. 2010)

1. How is dissolved iron transported from the Amur River basin to the Sea of Okhotsk and the Oyashio region?

Average annual fluxes of total and dissolved iron were estimated in various parts of the GFBF and the project confirmed the continuity of iron transportation from the land surfaces of the Amur River basin to the surface water of the Oyashio region (Fig. 8.2). Natural wetlands with gentle slopes located in the middle and lower parts of the Amur basin were the major source of dissolved iron from terrestrial zones to the Amur River.

In the upstream forested basin, dissolved iron in soil was mainly transported with dissolved organic carbon (DOC) rather than as Fe(II) and Fe(III). The riparian zone near the stream channel is an important source of iron owing to its wet and anaerobic condition, which increases the DOC concentration and dissolved iron in soil and groundwater.

In the natural wetlands, the dissolved Fe concentration is around 1 mg Fe L^{-1} in the surface water and much higher (sometimes more than 10 mg Fe L^{-1}) in soil interstitial waters, having a seasonal variation with maxima in summer. The dissolved Fe concentration observed for a number of rivers and agricultural drainage waters of the Sanjiang plain when not frozen has an average of approximately 1 mg Fe L^{-1} and varies considerably according to the condition of the watershed. Dissolved Fe is dominantly present as complex forms in soil water, river water and agricultural drainage water, in which humic substances play an important role in the transportation of iron as a complex ligand (Yoh et al. 2010).

As a result, $1.1 \pm 0.7 \times 10^{11}$ g/year of dissolved iron is transported to the estuarial area from the Amur River annually. Approximately 95% of the dissolved iron coagulates at Amur-Liman (the estuarial area) and Sakhalin Bay (Nagao et al. 2010). There are two pathways of iron transportation from the estuarial area to the Oyashio region: (1) surface transportation of total iron and (2) transportation with the North Pacific Intermediate Water (NPIW). The former supports primary production in the Sea of Okhotsk while the latter supports primary production in the Oyashio region. It is estimated that approximately $1.2–1.5 \times 10^8$ g/year of total iron is provided by the atmosphere and NPIW in the Oyashio region (Nakatsuka et al. 2010; Nishioka et al. 2007).

2. **To what extent does the supply of dissolved iron regulate primary production in the open waters?**

   It was found that of the iron used by the spring bloom in the Oyashio region, 40% is provided through the GFBF system and 60% is recycled through microbial processes (Nakatsuka et al. 2010). We are not yet certain about the relative importance of atmospherically derived iron to primary production in the Oyashio region because of its temporal sporadicity, spatial unevenness and insoluble nature (Matoba et al. 2010). In spite of this uncertainty, it is reasonably concluded that the iron controls phytoplankton growth in the Oyashio region because phytoplankton growth ceases under iron limitation at all high nitrate concentrations.

   It is yet uncertain to what extent the supply of dissolved iron regulates primary production in the open waters. This is due mainly to a lack of sufficient observational data on both annual changes in the dissolved iron flux and the biomass in the Oyashio region. To determine the role of the dissolved iron, we developed a three-dimensional coupled ecosystem physical model that includes the effect of iron on the Sea of Okhotsk (Okunishi et al. 2007). We hypothesized that four processes supply iron to sea water: atmospheric loading, input from the Amur River, dissolution from sediments, and regeneration by zooplankton and bacteria. We simulated 1 year, from 1 January 2001 to 31 December 2001. As a result, the model taking iron into account agreed well with the observation. However, we are not yet able to simulate the time series of the iron impact, since the model cannot simulate the NPIW, which we believe is the most important current in the transportation of riverine iron from the Amur River (see Chapter 9).

3. **How do land surface disturbances affect material circulation in the Amur-Okhotsk system?**

   The impact of land-use change on iron discharge was studied in experimental plots of upland fields and paddy fields on the Sanjiang plain, which were converted from natural wetlands several decades ago (Yoh et al. 2010; Fig. 8.3). Soil in the upland fields was found to remain in an oxidized condition throughout the year, implying the absence of iron discharge. In paddy fields, surface water and soil water had dissolved Fe concentrations somewhat lower than those of natural wetlands, but importantly, the controlled water discharge due to agricultural management is considered to largely lower the iron discharge.

   Paddy fields on the Sanjiang plain are irrigated with ground water in most cases. The strikingly high concentration of dissolved iron (largely in the form of Fe^{2+}) might indicate an additional iron source. However, elevated contents of

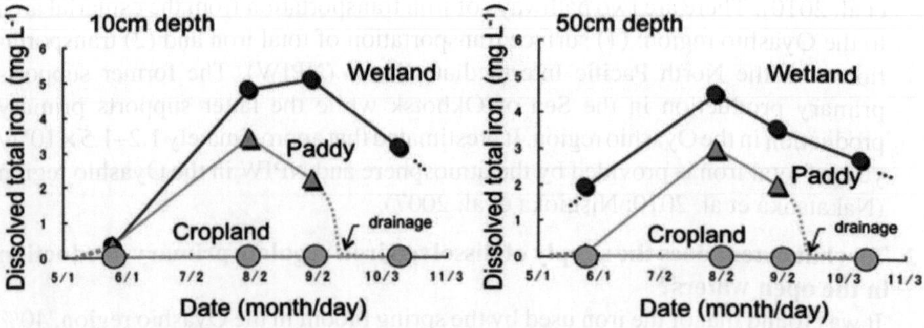

Fig. 8.3 Seasonal changes of dissolved iron concentration in soil solutions of three land-use types in Sanjiang Plain (Yoh et al. 2010)

amorphous iron oxides in the upper soil layer in paddy fields were found to adequately account for the calculated total amount of iron supplied by the irrigation of ground water since the rice paddy conversion on the Sanjiang plain, suggesting an almost complete retention of iron added by the ground water. Considering the irrigation and the controlled water discharge described above, it is concluded that iron discharge may be much less for paddy fields than for natural wetlands.

Monitoring data indicate that the concentration of dissolved iron in the Naoli River, which runs across the Sanjiang plain, has been consistently decreasing in recent decades (Yan et al. 2010). The observation of a peat layer, except in hilly areas, suggests a predominance of wetlands on the Sanjiang plain before the agricultural activities were started. However, a survey of the ground water table demonstrated that the current ground water levels were greatly lowered in most regions owing to reclamation by water drainage. It is likely that the land previously dominated by wetlands has been becoming steadily drier on the Sanjiang plain, which has reduced the Fe discharge as mentioned above.

Land-use and historical changes in the Amur River basin were visualized by various temporal and spatial mappings. We compiled land-use maps for both the 1930s and 2000 for the whole Amur River basin (Ganzey et al. 2010). Changes in the most recent 19 years were analyzed using Pathfinder AVHRR Land datasets and satellite remote-sensing techniques (Murooka et al. 2007). The results show significant changes on the Sanjiang plain in which approximately 10,000 km^2 of wetland was reclaimed as paddy fields from 1980 to 2000. Aerial changes of Russian forest were not significant but the quality of the forest is considered to be deteriorating mainly owing to frequent forest fires and poor management.

Such land-use changes were caused by various factors. According to analyses of the underlying causes of the degradation of forest resources in Khabarovsk Krai and the current state of forest management policy, forest fires and poor management systems are identified as the major causes of forest degradation (Kakizawa et al. 2010). The rapid increase in timber exports to China and poor

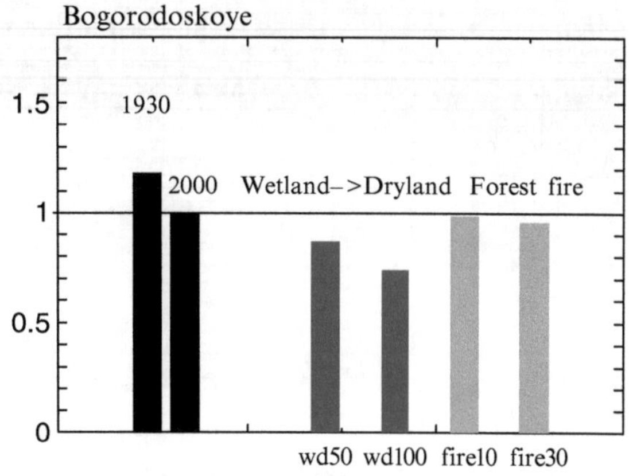

Fig. 8.4 Simulated results of land cover conversion (wd50: 50% converted, wd100: 100% converted, fire10: forest fire area is 10%, fire30: forest fire area is 30%) effect on dissolved iron productivity of the basin (Ohnishi et al. 2010)

forest policy are considered to accelerate forest degradation. On the Sanjiang plain, there was rapid development of paddy fields in accordance with governmental policy. Farm management has improved, but a lack of water has become a serious issue and the excessive pumping of ground water has caused the rapid lowering of the ground water table on the Sanjiang plain.

4. **How will human activity impact the system in the future?**
We developed a numerical hydro-geochemical model with special emphasis on iron dynamics for the Amur River basin (Onishi et al. 2010: see Chapter 9). The accuracy of the calculated discharge and dissolved iron concentration are sufficient at a time resolution of one month during the period from 1980 to 1990. Using the model, the effect of land cover change on dissolved iron productivity was evaluated. The results of numerical experiments suggest that 50% conversion of remaining wetlands to agricultural lands might decrease the dissolved iron flux by more than 10% (Fig. 8.4).

5. **How can we conserve this transboundary system?**
The key problem in conservation is how to establish a multilateral cooperative framework for the GFBF system. There have already been some bilateral frameworks, including the formal joint-monitoring program between China and Russia after the Songhua River accident involving a petrochemical company in the Chinese province of Jiling in 2005, and the cooperative program on the research, conservation and sustainable use of the ecosystems in the Sea of Okhotsk signed by Russia and Japan in 2009. However, there has been no multilateral governmental framework concerning the GFBF system. At this stage, joint-monitoring, data exchange and mutual communication at an academic level are necessary as a starting point for the protection of the GFBF system. For this purpose, we established the Amur Okhotsk Consortium as a multinational academic network to discuss the conservation and sustainable use of the GFBF (Hanamatsu et al. 2010; Fig. 8.5: see Chapter 12). The network can

Fig. 8.5 The *Amur Okhotsk Consortium* was established in 8 November 2009 on the occasion of the International Symposium on "Environmental Conservation of the Sea of Okhotsk: Cooperation between Japan, China and Russia"

be thought of as comprising "epistemic communities"; Peter Haas proposed that such networks of knowledge-based experts could help states identify the interests of these communities, frame issues for collective debate, propose specific policies, and identify salient points for negotiations. Our attempt is motivated by the history of the environmental protection of the Baltic Sea from marine pollution for over 30 years (see Chapter 17).

On the other hand, we have analysed existing international and domestic laws and policies that seem to be applicable for the conservation of the GFBF system. A future conservation framework would incorporate them as useful components. The results show that while environmental factors in GFBF have already been partially regulated by international and national laws and policies, these management regimes have been established and implemented independently, and they sometimes overlap or conflict; therefore, they are not adequate for the conservation of the whole GFBF system. We conclude that it is important to coordinate and strengthen existing laws and policies in an integrated manner to manage this system consistently and effectively (Fig. 8.6: see also Chapter 12).

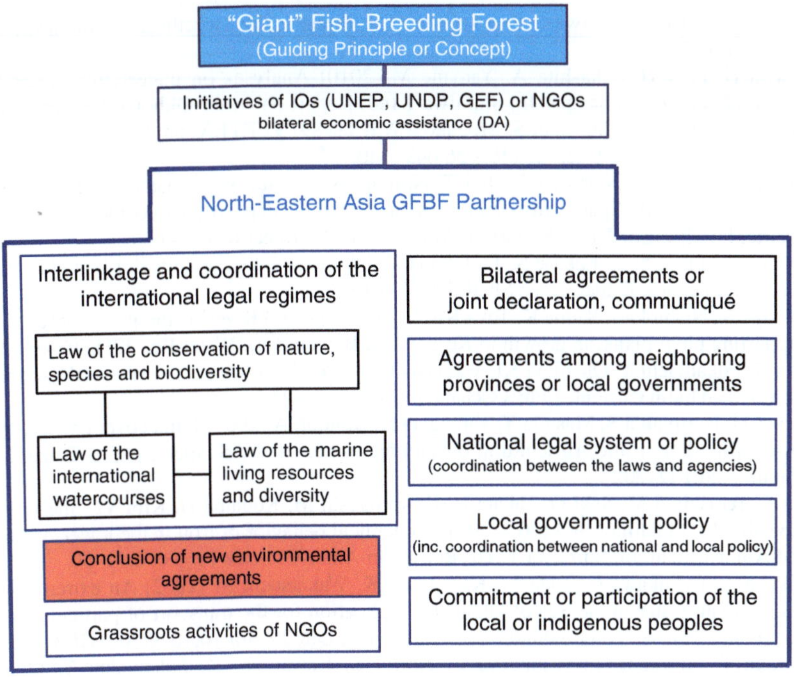

Fig. 8.6 Structure of the northeastern Asia GFBF partnership (Hanamatsu et al. 2010)

Acknowledgements I would like to thank all of the project member for their tireless efforts in achieving the tasks of the project. I am especially indebted to my colleagues from China and Russia for their cooperation and understanding in conducting the project. Special thanks are given to 10 group leaders of the Amur-Okhotsk Project and the project secretary, Ms. Tamaki Kawaguchi, who supported me from the very beginning to the end of the project. The philosophy of the Amur Okhotsk Project was developed under the guidance of Research Institute for Humanity and Nature (RIHN) and I would like to express my sincere gratitude to the late Director of RIHN, Toshihiko Hidaka and the present Director of RIHN, Narifumi Tachimoto.

References

Ganzey SS (2005) Transboundary geo-systems in the south of the Russian far east and in northeast China. Pacific Institute of Geography & Research Institute for Humanity and Nature, Dalnauka, Vladivostok, 235p

Ganzey SS, Ermoshin VV, Mishina N (2010) The landscape changes after 1930 using two kinds of land use maps (1930 and 2000). In: Shiraiwa T (ed) Report on Amur-Okhotsk Project 6:251–262

Hanamatsu Y, Horiguchi T, Endo T, Abe K (2010) The legal, political situations and a future conservation strategy of the giant fish-breeding forest. In: Shiraiwa T (ed) Report on Amur-Okhotsk Project 6:87–104

Hatakeyama S (1994) Moriwa umino koibito (The forest is the sweetheart of the sea), Hokuto Syuppan, 192p

Kakizawa H, Park H, Sakashita A, Yamane M (2010) Analyses on underlying causes behind land-use/land-cover changes. In: Shiraiwa T (ed) Report on Amur-Okhotsk Project 6: 63–70

Martin JH, Gordon RM, Fitzwater S, Broenkow WW (1989) VERTEX: phytoplankton/iron studies in the Gulf of Alaska. Deep-Sea Res 36:649–680

Matoba S, Minami H, Nishioka J, Ono T, Nomura M, Narita Y, Uematsu M, Muravyev Y.D, Shiraiwa T (2010) Spatial distribution of air-borne Fe deposition into the Northern North Pacific. In: Shiraiwa T (ed) Report on Amur-Okhotsk Project 6: 75–82

Matsunaga K, Igarashi K, Fukase S, Tsubota H (1984) Behavior of organically-bound iron in seawater of estuaries. Estuar Coast Shelf S 18(6):615–622

Matsunaga K, Nishioka J, Kuma K, Toya KA, Suzuki Y (1998) Riverine input of bioavailable iron supporting phytoplankton growth in Kesennuma Bay (Japan). Water Res 32(11):3436–3442

Mizuta G, Fukamachi Y, Ohshima KI, Wakatsuchi M (2003) Structure and seasonal variability of the East Sakhalin Current. J Phys Oceanogr 33:2430–2445

Murooka M, Haruyama S, Masuda Y, Yamagata K, Kondoh A (2007) Land cover change detected by satellite data in the Agricultural development area of the Sanjiang plain, China. J Rural Planning 26:197–202

Nagao S, Terashima M, Seki O, Takata H, Kawahigashi M, Kodama H, Kim VI, Shesterkin VP, Levshina IS, Makhinov AN (2010) Biogeochemical behavior of iron in the lower Amur River and Amur-Liman. In: Shiraiwa T (ed) Report on Amur-Okhotsk Project 6:41–50

Nakatsuka T, Yoshikawa C, Toda M, Kawamura K, Wakatsuchi M (2002) An extremely turbid intermediate water in the Sea of Okhotsk: Implication for the transport of particulate organic carbon in a seasonally ice-bound sea. Geophys Res Lett 29. doi:10.1029/2001GL014029

Nakatsuka T, Fujimune T, Yoshikawa C, Noriki S, Kawamura K, Fukamachi Y, Mizuta G, Wakatsuchi M (2004) Biogenic and lithogenic particle flux in the western region of the Sea of Okhotsk: implications for lateral material transport and biological productivity. J Geophys Res 109(C09):S13. doi:10.1029/2003JC001908

Nakatsuka T, Nishioka J, Suzuki K and all members of group 2 (2010) Iron transport processes and their impacts on primary productivity in the Sea of Okhotsk. In: Shiraiwa T (ed) Report on Amur-Okhotsk Project 6:31–40

Nishioka J, Ono T, Saito H, Nakatsuka T, Takeda S, Yoshimura T, Suzuki K, Kuma K, Nakabayashi S, Tsumune D, Mitsudera H, Johnson WK, Tsuda A et al (2007) Iron input into the western subarctic Pacific, importance of iron export from the Sea of Okhotsk. J Geophys Res 112:C10012. doi:10.1029/2006JC004055

Ohshima KI, Wakatsuchi M, Fukamachi Y, Mizuta G (2002) Near-surface circulation and tidal currents of the Okhotsk Sea observed with the satellite-tracked drifters. J Geophys Res 107:3195. doi:10.1029/2001JC001005

Okunishi T, Kishi MJ, Ono Y, Yamashita T (2007) A lower trophic ecosystem model including iron effects in the Okhotsk Sea. Cont Shelf Res 27(16):2080–2098

Onishi T, Shibata H, Yoh M, Nagao S, Park H, Shamov VV. (2010) Evaluation of land cover change impacts on dissolved iron flux of the Amur River. In: Shiraiwa T (ed) Report on Amur-Okhotsk Project 6:213–223

Simonov E.A, Dahmer T.D. (2008) Amur-Heilong River Basin Reader, WWF, Ecosystems Ltd, 426p

Takahashi T, Sutherland SC, Sweeney C, Poisson A, Metzl N, Tilbrook B, Bates N, Wanninkhof R, Feely R, Sabine C, Olafsson JA, Nojiri Y (2002) Global sea–air CO_2 flux based on climatological surface ocean pCO2, and seasonal biological and temperature effects. Deep Sea Res Part II 49(9–10):1601–1622

Tsuda A et al (2003) A mesoscale iron enrichment in the western subarctic Pacific induces a large centric diatom bloom. Science 300(5621):958–961

UNEP (2006) The Songhua River spill China, December 2005, Field Mission Report, UNEP, 26p

Yaginuma T (1999) Moriwa subete uotsukirin (All forest works as fish-breeding forest), Hokuto Syuppan, 246p
Yamashita Y, Tanaka M (eds) (2008) Linkage of forests, rivers and coasts and biological production in estuarine and coastal waters, Kouseisya-Kouseikaku, 147p
Yan B, Zhang B, Yoh M, Pan X (2010) Concentration and species of dissolved iron in waters in Sanjiang plain, China. In: Shiraiwa T (ed) Report on Amur-Okhotsk Project 6:183–194
Yoh M, Shibata H, Onishi T, Kawahigashi M, Guo Y, Yamagata K, Shamov VV, Levshina SI, Novorotskaya A, Matyushkina L, Yan B, Wang D, Pan X, Zhang B, Chen X, Huang B, Chi G, Shi H, Xu X, Zhang K, Cai T, Sheng H (2010) Iron dynamics in terrestrial ecosystems in the Amur River basin. In: Shiraiwa T (ed) Report on Amur-Okhotsk Project 6:51–62

Chapter 9
Numerical Simulation of Dissolved Iron Production and Transport in the Amur River and the Sea of Okhotsk

Takeo Onishi, Humio Mitsudera, and Keisuke Uchimoto

Abstract Wetlands in the Amur River basin play an important role in producing dissolved iron. A significant part of this dissolved iron is removed at the estuary zone. However, the mouth of the river is coincidentally located over the northwestern shelf where dense shelf water (DSW) is vigorously produced. Therefore, dissolved iron from the Amur River should be transported by DSW to the Sea of Okhotsk where iron is the limiting factor in primary production. During the last century, large areas of wetland within the Amur River catchment have been cultivated, which, in turn, may have had a great impact on the production of dissolved iron. To assess the impact of this conversion in land cover on the production of dissolved iron—and thereby the primary production in the Sea of Okhotsk—we constructed a numerical model to simulate dissolved iron production and transport in the Amur River and the Sea of Okhotsk.

The developed model consists of terrestrial and marine components. The terrestrial part of the model successfully simulates the discharge and dissolved iron flux of the basin. The marine part of the model also effectively simulates realistic DSW, which flows along the coast of Sakhalin through an intermediate (200–500 m deep) layer, and then experiences strong tidal mixing adjacent to the Kuril Islands in the southern Okhotsk Sea. Thus, our model can simulate dissolved iron production and transport in the Amur River basin and the Sea of Okhotsk.

Using the terrestrial part of the model, we simulated the impact of various land cover changes on the dissolved iron productivity of the Amur River basin. Two typical land cover change scenarios were proposed as possible situations in the basin.

T. Onishi (✉)
River Basin Research Center, Gifu University, 1-1 Yanagido, Gifu 501-1193, Japan
e-mail: takeon@gifu-u.ac.jp

H. Mitsudera • K. Uchimoto
Institute of Low Temperature Science, Hokkaido University, Hokkaido, Japan

M. Taniguchi and T. Shiraiwa (eds.), *The Dilemma of Boundaries: Toward a New Concept of Catchment*, Global Environmental Studies,
DOI 10.1007/978-4-431-54035-9_9,

One involves the conversion of wetland to agricultural land (paddy fields and dry land); the other is forest fire. Results indicated that the conversion of wetlands to agricultural land has a significant impact on the dissolved iron flux. Further research is needed to determine how wetland decrease might affect primary production in the Sea of Okhotsk. This paper consists of two parts. The first part addresses the results of the terrestrial simulation, and the second part deals with the results of the marine simulation.

Keywords Amur river • Dense shelf water • Dissolved iron • Land cover change • Sea of Okhotsk • Wetland

9.1 Amur River

The Amur River, which rises in Mongolia, is one of the largest rivers crossing the boundary between China and Russia. The catchment area of the river is 2,050,057 km^2, the ninth largest river catchment in the world, and the total length of the river is 4,350 km. Therefore, a huge quantity of fresh water is supplied by the Amur River to the Sea of Okhotsk (Ogi et al. 2001).

The Sea of Okhotsk is one of the most biologically productive regions in the world and, not surprisingly, it supports a highly productive commercial fishery. Martin and Fitzwater (1988) found that iron abundance limits phytoplankton growth in the Northeast Pacific Ocean. Recent studies show that dissolved iron plays an important role in maintaining the biological productivity of the Sea of Okhotsk (Boyd et al. 2004), and it is highly possible that one of the most important sources of dissolved iron is fresh water from the Amur River (Ducklow et al. 2003; personal communication with Dr. Nishioka). Iron is an essential nutrient not only for the biological productivity of the Sea of Okhotsk but also for most biota. However, the production and transportation of dissolved iron through the terrestrial part of this ecosystem is not well understood. A complicating factor, especially in the last few decades, is that extensive regions of agricultural land have been reclaimed by drainage improvements to wetlands (Wang et al. 2004).

To investigate the mechanism of dissolved iron production and to evaluate the effect of land cover change, we have developed a simple hydrological model incorporating dissolved iron production to simulate discharge and the distribution of dissolved iron in the Amur River basin. In this paper, following the description of the structure of the model and the source of data, an attempt to validate the calculated runoffs and dissolved iron concentrations is made. Then, based on the constructed model, the effect of land cover conversion on dissolved iron productivity in the basin is evaluated by running numerical simulations.

9.2 Site Description

Figure 9.1 shows the outline of the study site. The main tributaries of the basin are Songhua (Chinese part, 535,232 km^2), Argun (298,361 km^2), Zeya (233,311 km^2), Silka (202,924 km^2), Ussuri (195,101 km^2). The average river bed slope from the river mouth to Khabarovsk calculated from DEM data, which in turn were constructed using SRTM (Shuttle Radar Tomography Mission) data, is about 1/25,000. Compared with other large continental rivers, the Amur River basin is clearly very flat.

The dominant land cover in the basin is mixed deciduous and coniferous forests. Dry land occupies the next most significant proportion of the land cover. Most of the dry land is located in the Songhua River basin and occupies about 40% of the area of that basin. The main crops there are maize, corn and rice (National Bureau of Statistics of China 1949–2004). Wetland covers about 7% of the total basin area—mostly along the main course of the Amur River. From on-site investigations and periodic measurements of iron concentration, forest and wetland are seen to possibly be producing dissolved iron. To simulate this dissolved iron production in the Amur River basin, we constructed a simple hydrological model that incorporates dissolved iron production using a simple technique.

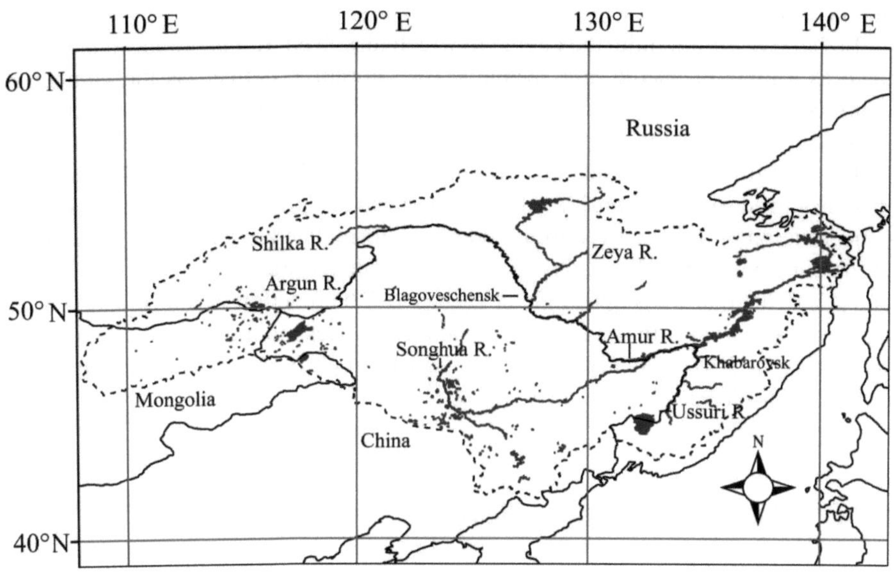

Fig. 9.1 Outline of the study site centered in northeast China

9.3 Materials and Methods

9.3.1 Model Structure

The constructed model used here is based on the TOPMODEL concept (Beven and Kirkby 1979). While TOPMODEL was originally developed to simulate runoff from a small scale catchment, it has also been used in global scale Land Surface Models such as MATSIRO (Takata et al. 2003). The model consists of two modules, one for dealing with the physical processes involved with runoff (TOP-RUNOFF), and the other for processes involved with the production of dissolved iron (TOP-FE). A schematic diagram of the model is shown in Fig. 9.2.

In the TOP-RUNOFF module, two processes are considered within the framework of the original TOPMODEL gridded structure. One is the inflow of surface water runoff from surrounding lands into the wetlands. This process was formulated using a simple gridding technique to account for the addition of surface runoff to the wetlands from sources other than paddy fields. The second process involved water management practices for paddy fields. This takes into account the overflows from paddies that occur when the ponding depth exceeds the prescribed threshold value PDc [m]. Thus, the modeling algorithm assumes that artificial drainage for agricultural purposes (e.g., mid-summer drainage) was not practiced in the basin. That said, some exceptions might exist in actual water management; the above formulation can be justified and modified by information obtained from field observations and farmers.

In the TOP-FE module, the degree to which dissolved iron is produced is formulated as a function of the time available for saturation, defined as the number of continuously saturated days. In the model, when both the root zone deficit and the saturation deficit of each grid reached zero, the grid was considered saturated. If the saturation duration time of a grid became larger than the threshold value SDc, then dissolved iron was considered to be produced at a prescribed constant rate. The concentration of the dissolved iron produced is formulated as a function of the topographic index $a/\tan\beta$. Here, a is defined as the drainage area of each calculation grid and β is defined as the slope angle of each grid. The function was formulated as an exponential curve with a range of parameters corresponding to land cover type (Onishi et al. 2010). Finally, the total runoff from each grid is routed using TRIP (Oki and Sud 1998).

9.3.2 Data and Parameter Setting

NCEP-DOE reanalysis-2 data with a spatial resolution of 2.5°×2.5° were utilized as climate forcing data. These data included short wave radiation, long wave radiation, specific humidity, wind speed, and air temperature. The daily precipitation rate was obtained from the APHRODITE dataset (Takashima et al. 2009) at a spatial

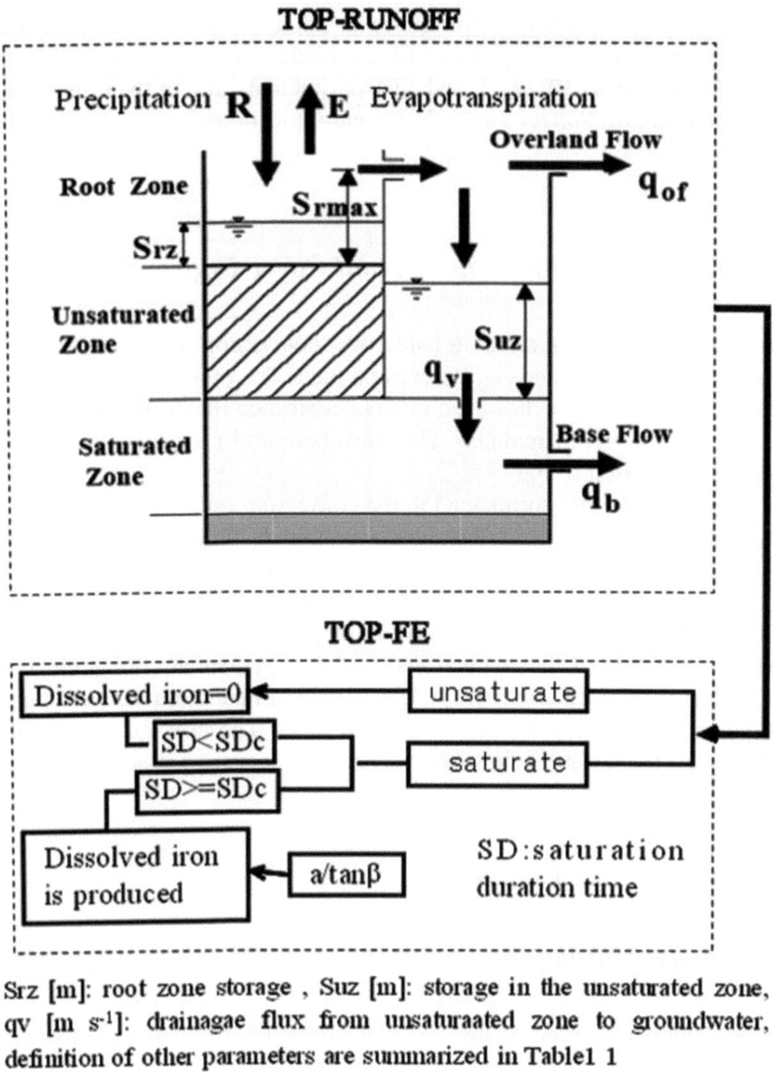

Fig. 9.2 Schematic diagram of the components of TOPMODEL (Beven and Kirkby 1979)

resolution of 0.5° × 0.5°. The evapotranspiration rate was calculated by the Penman–Monteith equation. The roughness length of vegetation, aerodynamic resistance and surface resistance of each land cover type were calculated using the formulations of Watanabe (1994). The leaf area index (LAI) and the vegetation height of each land cover type were based on measured values.

Daily discharge data and dissolved iron concentrations that were monitored by the Federal Service for Hydrometeorology and Environmental Monitoring (ROSHYDROMET) were used for model validation. Water collected for the analysis of dissolved iron concentration was filtered using a 0.45 mm filter, and analyzed by photometry. The grid size for the model was set as 1 km^2. The threshold value of SDc was set as 10 days based on experimental data for the iron oxide reduction rate (Roden and Wetzel 2002). Parameters for model calculation are listed in Table 9.1.

9.3.3 Experimental Design of Land Cover Conversion

Two typical scenarios that result in land-cover conversion were considered. One is the conversion from wetlands to agricultural land. The other is wild fire (excluding peat fires). In addition, the dissolved iron concentration under the land cover conditions of the 1930s were simulated. The conversion ratio for each scenario is summarized in the caption of Fig. 9.5.

In all cases, grids that correspond to the conversion ratio for each scenario were selected randomly and converted to agricultural lands or subjected to wild fire. The effect of wild fire on the productivity of dissolved iron is assumed to be negative. Based on comparative studies of dissolved iron productivity in natural and recently burnt forest, it was assumed that dissolved iron productivity after a forest fire would decrease to half that of the original natural forest (personal communication with Dr. H. Shibata). The time length used in all experiments was 10 years, using the climate data during the period from 1981 to 1990. The average annual dissolved iron flux under the different land cover conversion scenarios was compared.

9.4 Results and Discussion

9.4.1 Model Validation

Figure 9.3 compares observed and calculated monthly discharge at three observation stations along the main course of the Amur River. The results show fairly good agreement between calculated discharge and observed values, even though no calibration process was applied. Figure 9.4 compares the annual observed and calculated dissolved iron flux at the Khabarovsk station during the period between 1980 and 1990.

Note that the validation period and the time over which the land cover conditions were compiled are not the same. However, statistical data from Heilongjiang province (National Bureau of Statistics of China 1949–2004) suggest that expansion of agricultural land areas stabilized after 1980. Thus, it is assumed that land cover condition in the 1980s and in 2000 can be considered to be similar. Although the result shows non-negligible differences in some years, the calculated dissolved iron flux generally shows good agreement with observed values (Onishi 2007).

Table 9.1 List of parameters for model calculation

Symbol	Description	Unit	Resolution	Value	Source
Prescribed with horizontal distribution					
a/tanβ	Land cover type	–	100 m		Yermoshin et al. (2007)[a]
	Soil type	–	1°		ISLSCP II[b]
	Elevation	m	1000 m		SRTM
	Topographic index	m	1000 m		SRTM
Prescribed with land cover type					
LAI	Leaf area index	m^2/m^2	1000 m		Li et al. (2003), Liua et al. (2005), Scurlock et al. (2001)
	Surface conductance	m/s	1000 m		Kondo (1994)
	Aerodynamic conductance	m/s	1000 m		Kondo (1994)
Prescribed with soil type					
T_0	Saturated hydraulic conductivity	m/s	1°		ISLSCP II[b]
Prescribed as constant					
szm	Scaling parameter for runoff	m	–	0.001	–
SR_{max}	Maximum root zone deficit	m	–	0.01	–
t_d	Time constant for recharge to the saturated zone	m/h	–	0.1	–
chv	Channel routing velocity	m/s	–	0.5	–
rv	River routing velocity	m/s	–	0.5	–
Cw	Snow water retention capacity	–	–	0.1	–
T_s	Threshold temperature for 100% snow	K	–	2.0	Beven (2000)
T_r	Threshold temperature for 100% rain	K	–	4.5	Beven (2000)
T_m	Threshold temperature for snow melt	K	–	0.0	Beven (2000)
PD_c	Upper limit of ponding depth of paddy fields	m	–	0.1	–
SD_c	Threshold for starting of dissolved iron production	day	–	10	–

[a] Product of Amur-Okhotsk project
[b] Hall et al. (2005)

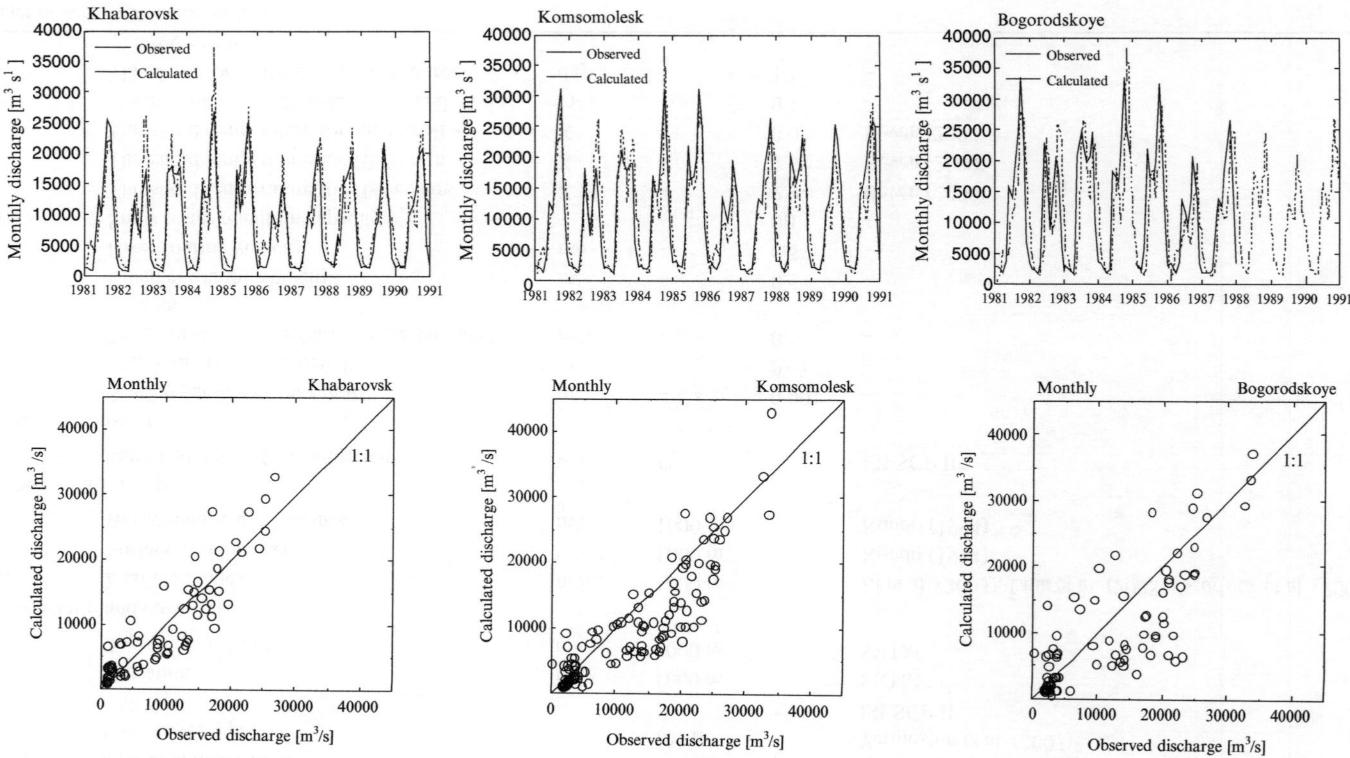

Fig. 9.3 Observed and calculated monthly discharge along the main course of the Amur River (at Khabarovsk, Komsomolesk, and Bogorodskoye)

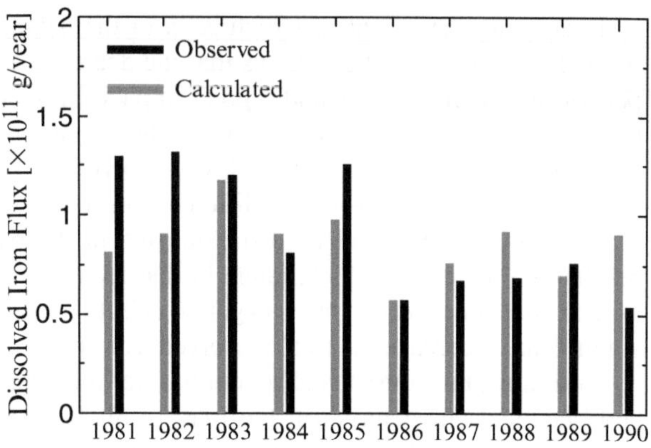

Fig. 9.4 Observed and calculated dissolved iron flux near the mouth (Bogorodskoye) of the Amur River basin

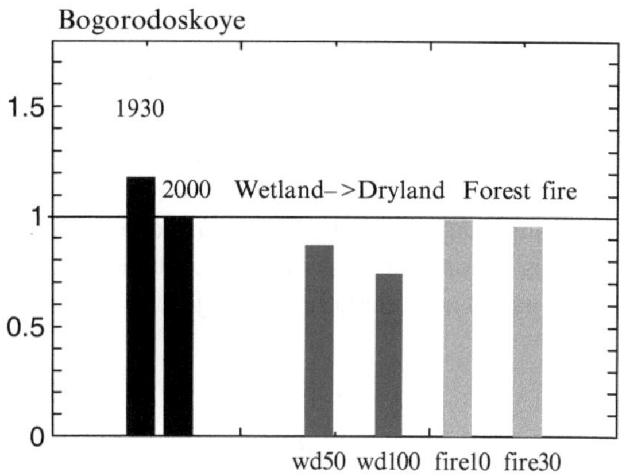

Fig. 9.5 Simulated results of the effect of land cover conversion on dissolved iron productivity in the basin (wd50: 50% converted, wd100: 100% converted, fire10: forest fire area is 10%, fire20: forest fire area is 20%). The vertical axis corresponds to relative increase/decrease compared with dissolved iron flux under the present land cover conditions

9.4.2 *Effect of Land Cover Conversion*

Figure 9.5 shows the dissolved iron flux as determined by numerical simulation near the mouth of the Amur River and several main tributaries (Zeya, Bureya, Songhua, and Ussuri).

The results suggest that the impact of agricultural development on dissolved iron flux is much greater than that of forest fires. This supports the idea that wetland conversion plays an important role in dissolved iron productivity, even though the areal extent of wetlands is not so large. Complete conversion of wetlands in the basin might result in a decrease in dissolved iron flux of approximately 20% compared with present conditions. Our experiments also indicate that dissolved iron flux under the land-cover conditions of the 1930s was more than 20% higher than under present land-cover conditions.

Examining the extent to which each of the tributaries contribute to the change in total dissolved iron flux shows that the decrease of wetlands within the Chinese part of the catchment has had a great impact on the dissolved iron productivity of the basin. In contrast, our experiments show that the influence of wild fire has not had much impact, because of the low dissolved iron productivity characteristics of forested regions. However, in addition to forest fires, there is some evidence that the scale of fires in peatland is also extensive in the basin. Whether such peat fires results in a distinguishable change in dissolved iron production is not clear. If we assume that peat fire also has a negative effect on dissolved iron productivity, it may significantly decrease the total dissolved iron flux.

Until now, the constructed model has not included the effect of slow changes in soil chemical characteristics that may result from the conversion of wetlands to agricultural lands. This effect might occur over a discrete time period due to gradual changes in soil chemical properties that occur after the conversion. However, our model formulates such an effect as being abrupt. Therefore, in observational data, we might expect some delay in the timing of chemical or physical responses to land-cover conversion.

Dissolved iron from the Amur River basin supplies a huge amount of iron to the Sea of Okhotsk. This iron is one of the most important factors supporting primary production in the sea. However, the effect of land-cover conversion on the primary production of the sea has not yet been resolved. This effect can be projected by using coupled hydrological and ocean circulation models. Coupling these two models beyond the boundary of land and ocean to simulate water movement seamlessly is the next challenging subject.

9.5 Sea of Okhotsk

The Sea of Okhotsk has been recognized as one of the primary sources of iron in the northwestern Pacific (Nishioka et al. 2007). Cold and dense water, referred to as dense shelf water (DSW), forms as a result of brine rejection from sea ice on the northwestern shelf in the Sea of Okhotsk. Iron, together with other chemical particles, is mixed into DSW by tidal mixing and turbulence (Nakatsuka et al. 2004). DSW is transported southward by the East Sakhalin Current in the intermediate layer off the coast of Sakhalin, and flows out to the Pacific through the Kuril Straits. Thus, iron is supplied to the northeastern Pacific, which supports, at least partly, the biogeochemical cycle in this region.

We are currently aiming to model such iron circulation via intermediate-layer circulation. We have constructed an Ocean General Circulation Model (OGCM), and investigated overturning circulation in the Sea of Okhotsk (Matsuda et al. 2009). As the first step in incorporating an iron model, we have conducted tracer experiments to validate the performance of the transport processes of the model, including realistic chlorofluorocarbon (CFC) input (Uchimoto et al. 2011). In this paper, we describe circulation in the OGCM and experiments with a simple tracer. Construction of the iron model is under development, and some results from a one-dimensional model are shown.

9.6 OGCM

The model used here is Iced COCO ver. 3.4 (Hasumi et al. 2004). This is an ice-ocean coupled model developed at the Center for Climate System Research, University of Tokyo. In the ocean model, the vertical coordinate system is a hybrid of sigma (between the sea surface and a depth of 31 m) and z level (below 31 m) coordinates. The partial step formulation is adopted for bottom topography (Adcroft et al. 1997). For the tracer equations, the advection schemes are the Quadratic Upstream Interpolation for Convective Kinematics with Estimate Streaming Terms (Leonard 1979) and the Uniformly Third-Order Polynominal Interpolation Algorithm (Leonard et al. 1993). Furthermore, isopycnal diffusion (Cox 1987), thickness diffusion (Gent et al. 1995), and the turbulence closure of Noh and Kim (1999) are used. The isopycnal and the thickness diffusion coefficients are 1.0106 and 3.0106 cm^2/s, respectively, and the background vertical viscosity and diffusion coefficients are 1.0 and 0.1 cm^2/s, respectively. In the ice model, the thermodynamic part is the zero layer model (Semtner 1976), and the dynamic part is the elastic-viscous-plastic formulation of Hunke and Dukowicz (1997) with a two-category thickness representation.

The model domain spans the region from 136°E to 179.5°W and from 39°N to 63.5°N. The horizontal resolution is 0.5° both in zonal and meridional directions. There are 51 levels in the vertical direction with thickness increasing from 1 m at the sea surface to 1000 m in the deepest layer.

The OGCM is forced at the sea surface by the daily mean climatology data set (wind stress, fresh water flux, radiations, wind speed, temperature, and humidity) from the Ocean Model Intercomparison Project (Röske 2001). The fresh water flux data consist of evaporation, precipitation and river runoff. Although the fresh water flux is much greater at the northern mouth of the Mamiya (Tartar) Straits than elsewhere throughout the year, probably owing to runoff from the Amur River, the runoff ought to drop or stop in winter because of freezing of the river. Therefore, we subtract the annual mean (which we regard as approximate river runoff) from the data at each grid north of 53°N and west of 142°E in winter (from 15th December to 15th April); the amount subtracted is evenly shared out to the rest of the days.

Temperature and salinity are restored to the World Ocean Atlas (WOA) data on the six grids from the boundaries, and the sea surface height is restored to the basin wide model outputs on the three grids from the boundaries. Temperature and salinity are also restored to the WOA at grids deeper than 2000 m. The sea surface salinity (SSS) is also restored to the WOA. From December to April, the SSS is not restored to the WOA in the northern half of the Sea of Okhotsk as the SSS around the northwestern part of the Sea of Okhotsk in the WOA is too low in winter, probably as a result of spurious effects in the Amur River runoff.

Although circulation in and around the Sea of Okhotsk is strongly affected by tidal mixings along the Kuril Islands, this OGCM does not include tidal effects. Therefore, we increased the vertical diffusion coefficients to represent tidal effects along the Kuril Islands, with the coefficients set at 500 cm^2/s at the bottom and decrease upward.

The OGCM is integrated for 116 years with constant temperature and salinity at all grids. The next section describes monthly means of the results in the last year of this integration.

9.7 Simulation of Okhotsk Sea Circulation

Figure 9.6 shows barotropic streamlines for winter and summer. Note that the cyclonic gyre in the center of the Sea of Okhotsk is strong in winter and weak in summer. In summer, in the southern part of the sea near the Kuril Islands, the values are positive and some anticyclonic circulation is seen. This circulation and its seasonal variations are consistent with previous studies (e.g. Ohshima et al. 2004; Uchimoto et al. 2007).

Figure 9.7 shows an ice concentration map for February. This shape of ice distribution is similar to that observed by satellites (e.g. Ohshima et al. 2006).

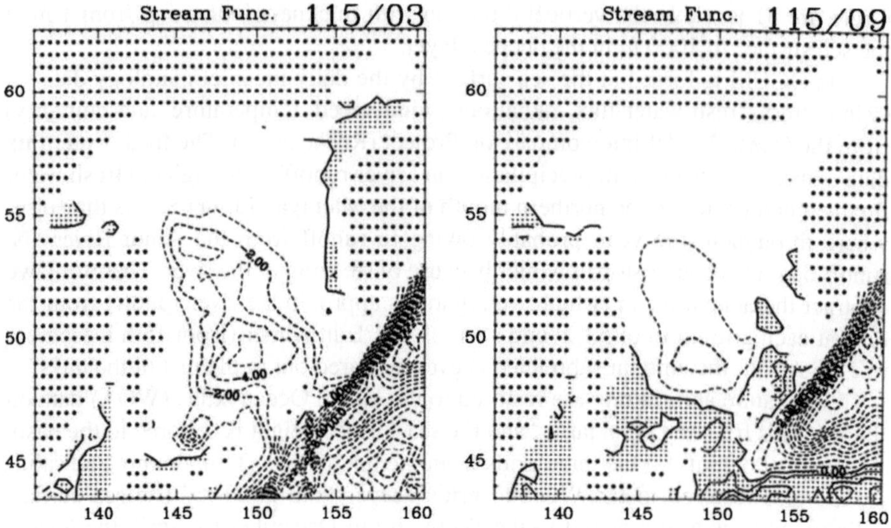

Fig. 9.6 Barotropic streamlines in March (*left*) and September (*right*). *Shading region* denotes positive values of streamlines. Contour interval is 1 Sv

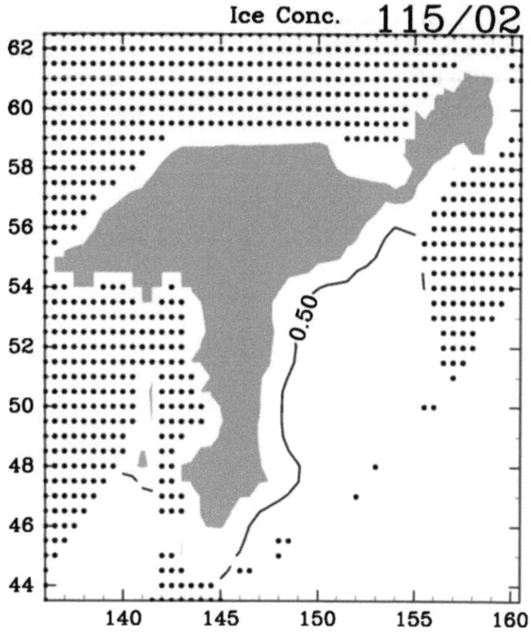

Fig. 9.7 Ice concentration map for February. *Shaded region* is where the concentration is larger than 0.95. The contour denotes a concentration of 0.5

We should note that the concentration is somewhat low (less than 0.95) along the northern coast. This low concentration is associated with coastal polynyas, where sea ice is produced continually and therefore brine is rejected efficiently. Since the OGCM represents ice distribution fairly well, DSW is also expected to be represented well.

Next, several properties on the $26.8\sigma_\theta$ surface are shown. The $26.8\sigma_\theta$ surface is a typical surface of the intermediate layer. Figure 9.8 shows its depth. The model represents a general resemblance to the climatological study of Itoh et al. (2003), although the isopycnal surface of this model is somewhat shallower. One of the salient climatological features is that the isopycnal is deep (~250 m) in the eastern basin, while it is shallow at the western boundary, which is well represented in the simulation.

The $26.8\sigma_\theta$ surface in the OGCM outcrops in winter around the northwestern shelf (not shown). Cold water extends southward from the shelf along Sakhalin Island (Fig. 9.9). In the central and eastern parts of the sea, the potential temperature is relatively high because warm Pacific water enters there.

In summary, the OGCM represents the circulation and properties of the $26.8\sigma_\theta$ in the Sea of Okhotsk quite well in spite of its coarse resolution.

Fig. 9.8 Depth of the 26.8 σ_θ surface in September. Contour interval is 15 m

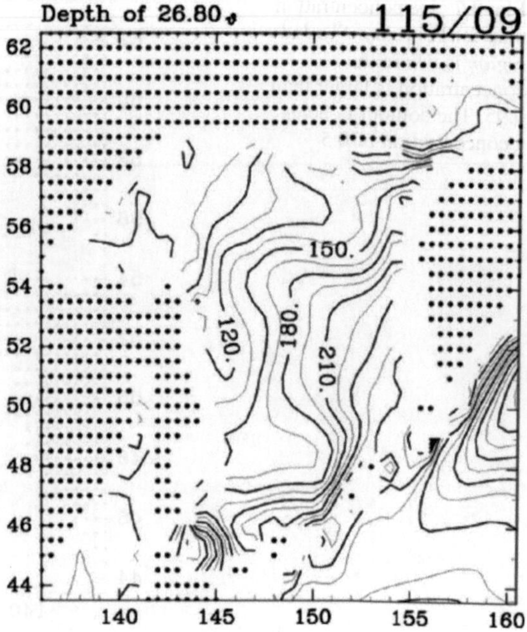

Fig. 9.9 Temperature on the 26.8σ_θ surface in September. Contour interval is 0.4°C. Regions where temperature is less than 0°C are *shaded*

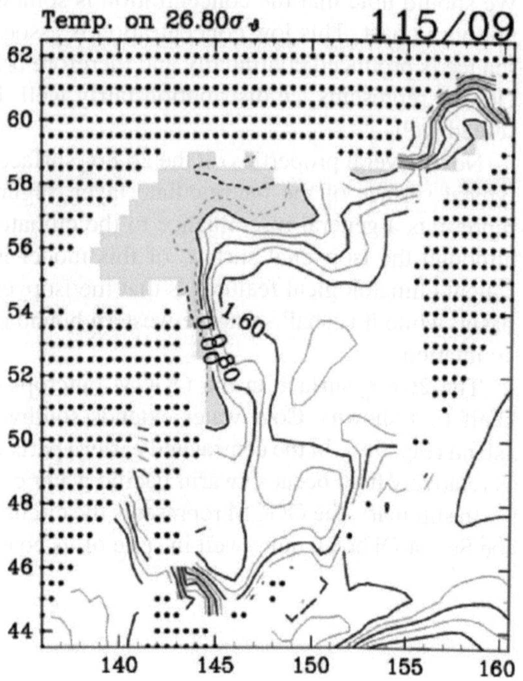

9.8 Tracer Experiments

To map the path of water that is ventilated lifted as a result of cooling and brine rejection around the northwestern shelf, we conducted tracer experiments, where passive tracers are restored to 1 from January to April. Figure 9.10 displays the tracer distribution on the 26.8σ_θ surface 11 years after the commencement of the experiments. The tracer is injected at the sea surface above the northwest shelf. It flows down to a depth of about 300 m, north of Sakhalin then moves further southward along Sakhalin Island toward the Kuril Islands. Some of the tracer recirculates within the Sea of Okhotsk, while most of it flows out into the Pacific through the Kuril Straits. The tracer finally flows to the east along the Oyashio front and distributes throughout the entire western Pacific.

These features, which pertain to material transport in the intermediate layer, are consistent with those of iron circulation deduced by Nishioka et al. (2007) from observations. We have checked the model performance concerning material circulations by simulating the CFC12 concentration (Uchimoto et al. 2011) and found that the model quantitatively represents CFC12 concentrations observed by Yamamoto-Kawai et al. (2004).

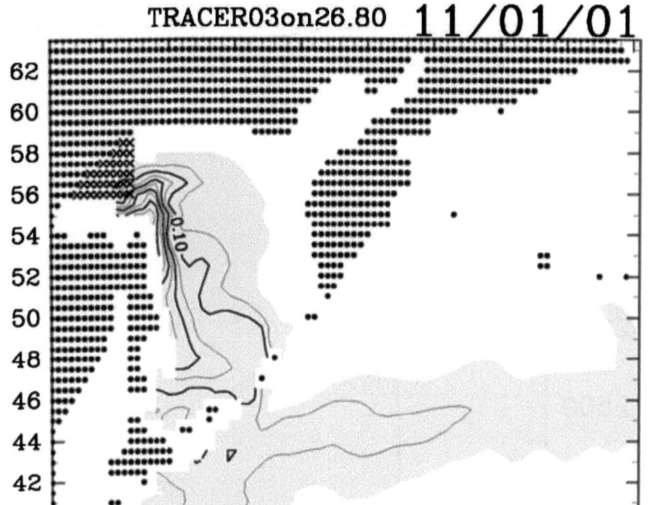

Fig. 9.10 Tracer concentration on the 26.8σ_θ surface 11 years from the start. Tracers are injected from January to April at the sea surface in the northwestern shelf denoted by crosses. Contour interval is 0.05. Regions where tracer concentration is larger than 0.1 are *shaded*

9.9 Iron Model

The iron model is based on the coupled oceanic iron and phosphorus cycle (Parekh et al. 2005). The iron parameterization includes scavenging and complexation with a ligand. Export production, which is described by a prescribed function, is limited by light, phosphate and iron.

Figure 9.11 represents a solution of a one-dimensional version of the iron-phosphorus model. The solution indeed represents the iron profile quite well. It remains for us to incorporate this iron model within the 3-D material circulation model described previously. This work is under development.

9.10 Concluding Remarks

In this article, several results from our OGCM that covers the Sea of Okhotsk and the northwestern Pacific have been described. In spite of its coarse resolution, the OGCM can represent circulation and active tracer distributions in the intermediate

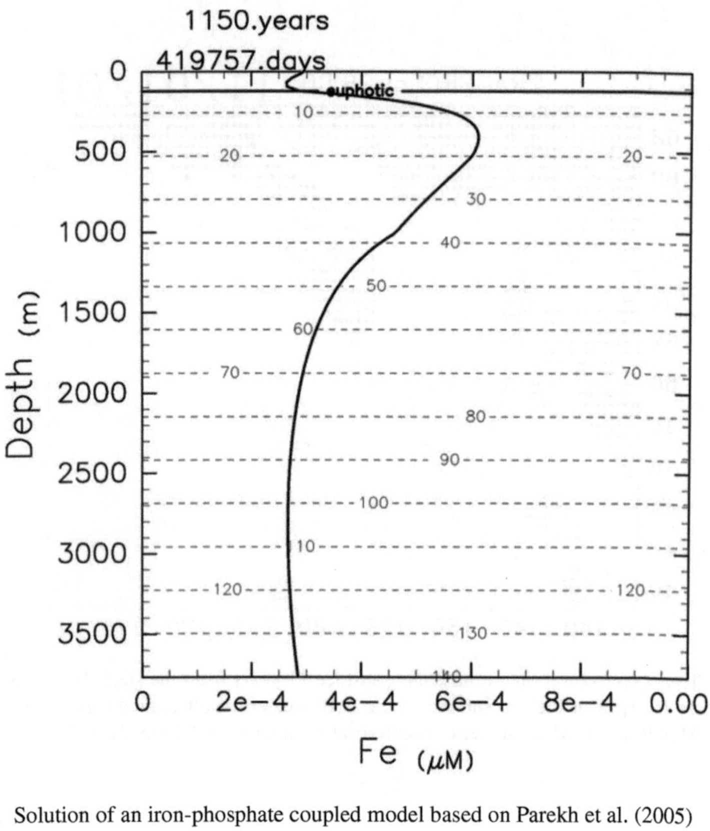

Fig. 9.11 Solution of an iron-phosphate coupled model based on Parekh et al. (2005)

layer reasonably well. Furthermore, passive tracer experiments show that dense and cold water (DSW) produced on the northwestern shelf in the Sea of Okhotsk flows southward along Sakhalin in the intermediate layer, and is finally distributed in the entire western North Pacific. This feature of iron distribution is similar to those observed in the Amur-Okhotsk project. We are now coupling an iron model to the OGCM to further investigate iron circulation.

Acknowledgments The terrestrial system work was undertaken under the framework of an international project entitled 'Human activities in Northeastern Asia and their impact on the biological productivity in North Pacific Ocean' funded by the Research Institute for Humanity and Nature (RIHN). The authors express their thanks to Global Soil Data Task (2000), the IGBP, ORNL DAAC and ISLSCP Initiative II for providing soil characteristic data. We deeply thank all the staff in ROSHYDROMET, and in particular G. Alexander for his special collaboration on this work. The ocean modeling work was benefitted by discussions with Dr. Nishioka and Dr. Nakamura of Institute of Low Temperature Science, Hokkaido University. We also thank for NEDO for their support.

References

Adcroft A, Hill C, Marshall J (1997) Representation of topography by shaved cells in a height coordinate ocean model. Mon Wea Rev 125:2293–2315

Beven KJ (2000) Rainfall runoff modelling: the primer. Wiley, New York, 372p

Beven KJ, Kirkby MJ (1979) A physically based variable contributing area model of basin hydrology. Hydrolog Sci Bull 24:43–69

Boyd PW, Law CS, Wong CS, Nojiri Y et al (2004) The decline and fate of an iron-induced subarctic plankton bloom. Nature 428:549–553

Cox MD (1987) Isopycnal diffusion in a z-coordinate ocean model. Ocean Model 74:1–5

Ducklow HW, Oliver JL, Smith WO Jr (2003) The role of iron as a limiting nutrient for marine plankton processes. In: Melillo JM, Field CB, Moldan B (eds) Interaction of the major biogeochemical cycles. Island Press, Washington, pp 295–310

Gent PR, Willebrand J, McDougall TJ, McWilliams JC (1995) Parameterizing eddy induced tracer transports in ocean circulation models. J Phys Oceanogr 25:463–474

Global Soil Data Task (2000) Global gridded surfaces of selected soil characteristics (IGBPDIS). International Geosphere–Biosphere Programme—Data and Information Services

Hall FG, Collatz G, Los S, Brown de Colstoun E, Landis D (2005) (eds) ISLSCP Initiative II. NASA. DVD/CD-ROM, NASA

Hasumi H, Oka A, Komuro Y (2004) A study on the polar ocean freshwater budget and the global thermohaline circulation using the CCSR sea ice-ocean model (Iced-COCO). CGER's Supercomputer Activity Report 11–2002:11–20

Hunke EC, Dukowicz JK (1997) An elastic-viscous-plastic model for sea ice dynamics. J Phys Oceanogr 27:1849–1867

Itoh M, Ohshima KI, Wakatsuchi M (2003) Distribution and formation of Okhotsk Sea intermediate water: an analysis of isopycnal climatological data. J Geophys Res 108(C8). 3258. doi:10.1029/2002JC001590

Kondo J (1994) Meteorology of water environment: water and heat budget at the surface. Asakura Press, 368p (in Japanese)

Leonard BP (1979) A stable and accurate convective modelling procedure based on quadratic upstream interpolation. Comput Meth Appl Mech Eng 19:59–98

Leonard BP, MacVean MK, Lock AP (1993) Positivity-preserving numerical schemes for multidimensional advection. NASA Tech Memo 106055:ICOMP-93-05

Li Y, Cui J, Zhang T, Zhao H (2003) Measurement of evapotranspiration of irrigated spring wheat and maize in a semi-arid region of north China. Agr Water Manag 61:1–12

Liua X, Jina J, Herbert SJ, Zhanga Q, Wang G (2005) Yield components, dry matter, LAI and LAD of soybeans in northeast China. Field Crop Res 93:85–93

Martin JH, Fitzwater SE (1988) Iron deficiency limits phytoplankton in the northeast Pacific subarctic. Nature 311:341–343

Matsuda J, Mitsudera H, Nakamura T, Uchimoto K, Nakanowatari T, Ebuchi N (2009) Wind and buoyancy driven intermediate-layer overturning in the Sea of Okhotsk. Deep-Sea Res I 56:1401–1418

Nakatsuka T, Fujimune T, Yoshikawa C, Noriki S, Kawamura K, Fukamachi Y, Mizuta G, Wakatsuchi M (2004) Biogenic and lithogenic particle fluxes in the western region of the Sea of Okhotsk: implications for lateral material transport and biological productivity. J Geophys Res 109:C09S13. doi:10.1029/2003JC001908

National Bureau of Statistics of China (ed) (1949–2004) Heilongjiang statistical year book. China Statistics Press. Beijing

Nishioka J, Ono T, Saito H, Nakatsuka T, Takeda S, Yoshimura T, Suzuki K, Kuma K, Nakabayashi S, Tsumune D, Mitsudera H, Johnson WK, Tsuda A (2007) Iron supply to the western subarctic Pacific: importance of iron export from the Sea of Okhotsk. J Geophys Res 112:C10012. doi:10.1029/2006JC04055

Noh Y, Kim J (1999) Simulations of temperature and turbulence structure of the oceanic boundary layer with the improbed near-surface process. J Geophys Res 104:15621–15634

Ogi M, Tachibana Y, Nishio F, Danchenkov MA (2001) Does the fresh water supply from the Amur River flowing into the Sea of Okhotsk affect sea ice formation? J Meteorol Soc Jpn 79:123–129

Ohshima KI, Simizu D, Itoh M, Mizuta G, Fukamachi Y, Riser SC, Wakatsuchi M (2004) Sverdrup balance and the cyclonic gyre in the Sea of Okhotsk. J Phys Oceanogr 34:513–525

Ohshima KI, Nihashi S, Hashiya E, Watanabe T (2006) Interannual variability of sea ice area in the Sea of Okhotsk: importance of surface heat flux in fall. J Meteorol Soc Jpn 84:907–919

Oki T, Sud YC (1998) Design of total runoff integrating pathways (TRIP)—a global river channel network. Earth Interact 1(2):1–37

Onishi T (2007) Runoff properties of the Amur River and the construction of the hydrological model incorporating dissolved iron transport. Annual Report of Amur-Okhotsk Project No. 4, pp 201–206

Onishi T, Muneoki Y, Shibata H, Nagao S, Kawahigashi M, Shamov VV (2010) Topography as a macroscopic index for the dissolved iron productivity of different land cover types in the Amur River Basin. Hydrol Res Let doi:10.3178/HRL.4.85

Parekh P, Follows MJ, Boyle EA (2005) Decoupling of iron and phosphate in the global ocean. Global Biogeochem Cycles 19:GB2020

Roden EE, Wetzel RG (2002) Kinetics of microbial Fe(III) oxide reduction in freshwater wetland sediments. Limnol Oceanogr 47:198–211

Röske F (2001) An atlas of surface fluxes based on the ECMWF re-analysis—A climatological dataset to force global ocean general circulation models. Max-Planck-Institut660 für Meteorologie Rep 323:31

Scurlock JMO, Asner GP, Gower ST (2001) Worldwide historical estimates of leaf area index, 1932–2000. ORNL technical memorandum ORNL/TM-2001/268. Oak Ridge National Laboratory, Oak Ridge

Semtner AJ Jr (1976) A model for the thermodynamic growth of sea ice in numerical investigations of climate. J Phys Oceanogr 6:379–389

Takashima H, Yatagai A, Kawamoto H, Arakawa O, Kamiguchi K (2009) Hydrological balance over northern Eurasia from gauge-based high-resolution daily precipitation data. In: Taniguchi M (ed) From headwaters to the ocean: hydrological change and watershed management. Taylor & Francis, London, pp 35–41

Takata K, Emori S, Watanabe T (2003) Development of the minimal advanced treatments of surface interaction and runoff. Global and Planetary Change. doi:10.1016/S0921-8181(03)00030-4

Uchimoto K, Mitsudera H, Ebuchi N, Miyazawa Y (2007) Anticyclonic eddy caused by the Soya warm current in an Okhotsk OGCM. J Oceanogr 63:379–391

Uchimoto K, Nakamura T, Nishioka J, Mitsudera H, Yamamoto-Kawai M, Misumi K, Tsumune D (2011) Simulations of chlorofluorocarbons in and around the Sea of Okhotsk: Effects of tidal mixing and brine rejection on the ventilation. J Geophys Res 116:C02034. doi: 10.1029/2010JC006487

Wang Z, Zhang S, Bai Z (2004) Effects of land use change on values of ecosystem services of Sanjiang Plain, China. China Environ Sci 24(1):125–128

Watanabe T (1994) Bulk parameterization for a vegetated surface and its application to a simulation of nocturnal drainage flow. Boundary-Layer Met 70:13–35

Yamamoto-Kawai M, Watanabe S, Tsunogai S, Wakatsuchi M (2004) Chlorofluorocarbons in the Sea of Okhotsk: ventilation of the intermediate water. J Geophys Res 109:C09S11. doi:10.1029/2003JC001919

Yermoshin VV, Ganzei SS, Murzin AV, Mishina NV, Kudryavtzeva EP (2007) Creation of GIS for Amur River basin: the basic geographical information. Annual Report of Amur-Okhotsk Project No. 4, pp 151–159

Chapter 10
Transportation of Marine-Derived Nutrients (MDN) onto Land by Anadromous Fish: A Survey with Reference to Pacific Salmon in the Russian Far East

Takeshi Murota

Abstract Land, ocean, and atmosphere are connected with each other through many pathways and/or material cycles. Rivers are the major corridor of nutrients transport, upstream and downstream, between them. While there have been many important studies in North America which demonstrate that anadromous fish, especially Pacific salmon (*Oncorhyncus* spp.), play a significant role in transporting marine-derived nutrients (MDN) onto terrestrial ecosystems, there have been few such salmon studies for the Russian rivers such as the Amur and the Anadyr Rivers. The purpose of this paper is to fill the gap of research between North America and Russia. Using data on escapement of adult Pacific salmon to spawning areas in the Russian Far East, the paper shows preliminary estimates as of how much marine-derived nitrogen (N) and phosphorus (P) are annually uploaded onto terrestrial ecosystems from the Northern Pacific. Through the concept of uo-tsuki-rin (fish breeding forest), terrestrial forests are known to nourish the ocean with iron and other life-support materials. Through a survey of past MDN studies, this paper points out the importance of life-support matters' transport in the opposite direction, i.e., from ocean to land.

Keywords Escapement data • Marine-derived nutrients (MDN) • Material cycle • Nutrients transport • Russian Far East • Uo-tsuki-rin

10.1 Introduction

Salmon rivers stink awfully in the spawning season. While wide spread odor of salmon carcasses is not pleasant for the human nose, it is a distinct sign of an ecologically healthy relationship between ocean and land in the Northern Pacific regions. The strong odor attracts wild mammals and birds to areas along the

T. Murota (✉)
Department of Economics, Doshisha University, Kyoto 602-8580, Japan
e-mail: tmurota@mail.doshisha.ac.jp

M. Taniguchi and T. Shiraiwa (eds.), *The Dilemma of Boundaries: Toward a New Concept of Catchment*, Global Environmental Studies,
DOI 10.1007/978-4-431-54035-9_10, © Springer Japan 2012

riverbeds. Such riversides are the seasonal restaurants which make marine-derived nutrients (MDN) available to wild animals in the forms of live adult salmon and their carcasses. MDN-dissolved water pools also serve as nursery grounds for many kinds of mosses and insect larvae. Salmon-fed animals drop excreta on the forest soil and thereby fertilize the catchment area.

There have been many important studies in North America which demonstrate that anadromous fish, especially Pacific salmon (*Oncorhyncus* spp.), play a significant role in transporting marine-derived nutrients (MDN) onto terrestrial ecosystems. Following the pioneering work of Juday et al. (1932), a number of analyses have quantified amounts of MDN and energy taken into the bodies of animals and plants or determined the structure nutrient transfer assumes in the catchment area if MDN are massively available (Cederholm et al. 1989; Bilby et al. 1996; Hilderbrand et al. 1996; Cederholm et al. 1999; Helfield and Naiman 2001; Bilby et al. 2003; Lessard et al. 2003; Zhang et al. 2003; Scheuerell et al. 2005; Schindler et al. 2005; Mertz and Moyle 2006; Brock et al. 2007; Uchiyama et al. 2008; Lessard and Merritt 2009, to cite a few).

In the Russian context, some pioneering studies of a similar nature appeared during the Soviet era, such as Krokhin (1957, 1967) work on sockeye salmon nursery lakes in Kamchatka. Such salmon studies stopped after the 1970s, despite there being many salmon rivers in the Russian Far East. Work by Nagasaka and colleagues (2006) has now restarted this research area. The paper, a result of joint research by Japanese and Russian scientists using a stable isotope approach, demonstrated the positive contribution of MDN to the growth of riparian vegetation along some rivers in Etorofu Island, the Northern Territory as a part of the Kuril Archipelago.

This paper aims firstly to provide a survey of existing MDN literatures, including the author's own, and secondly to extend the scope of the recent Nagasaka et al. (2006) study to the entire region of the Russian Far East (the continental area, Sakhalin Island, and the Kurils). Using data on escapement of adult Pacific salmon to spawning areas in the Russian Far East, the paper reviews my preliminary estimates in Murota (2004) of the amounts of marine-derived nitrogen (N) and phosphorus (P) which are annually uploaded onto terrestrial ecosystems from the Northern Pacific. This tentative result will then be compared with the case of British Columbia, Canada.

10.2 Importance of the MDN Study

If nutrients are preserved on land, it is easy for such nutrients flow to the ocean. There is the general tendency of materials available for living beings to be lost from terrestrial ecosystems to the oceans as a result of the universal law of gravity. No object on the earth, whether it is organic or inorganic, can be free from the traction force proportional to its mass due to the gravity law. This implies that all materials heavier than water eventually descend from high lands to the bottoms of the ocean.

If this process was uni-directional, however, lands would lose most of their nutrients and become totally barren while oceans would become overly eutrophicated. Since this obviously is not the case, there must be some mechanism which transports nutrients in the opposite direction (Murota 1998, pp. 123–124). From the viewpoint of material cycles at the global scale, Murota (2003) considered possible agents of transporting MDN onto terrestrial lands. Three major agents are identified: (1) fisheries by human hands, (2) guano occurrences through sea birds, and (3) spawning run of anadromous fish. Among these, (2) and (3) are largely supported by upwelling of deep sea water. The focus of this paper is on the amounts of MDN which are embodied in spawning salmon which are not caught by human beings and are counted as escapements in the official statistics of the Russian Far East.

10.3 Maximum Distance and Altitude of Spawning Run of Salmon

Some salmon spawn near the river mouth if the condition of riverbed spring, pebbles, rocks and others is suitable. On the other hand, some salmon go upstream much farther. To understand the deep connection between ocean and land, it is worthwhile to know the maximum distance and altitude of spawning salmon run from the sea coast in the northern hemisphere. Figure 10.1 is the reproduction of such data in Murota (2003, p. 24) for Pacific salmon (*Oncorhyncus* spp.) and Atlantic salmon (*Salmo salar*). This figure was made by the author's literature and map survey of salmon run in history and present.

According to Mathews (1986) and Jarman (1972), some group of Chinook salmon (*Oncoryncus tshawytscha*) travel as far as 3,000 km or more from the mouth of the Yukon River (quoted in Murota 2003, p. 23). The points C and D show such maxi-

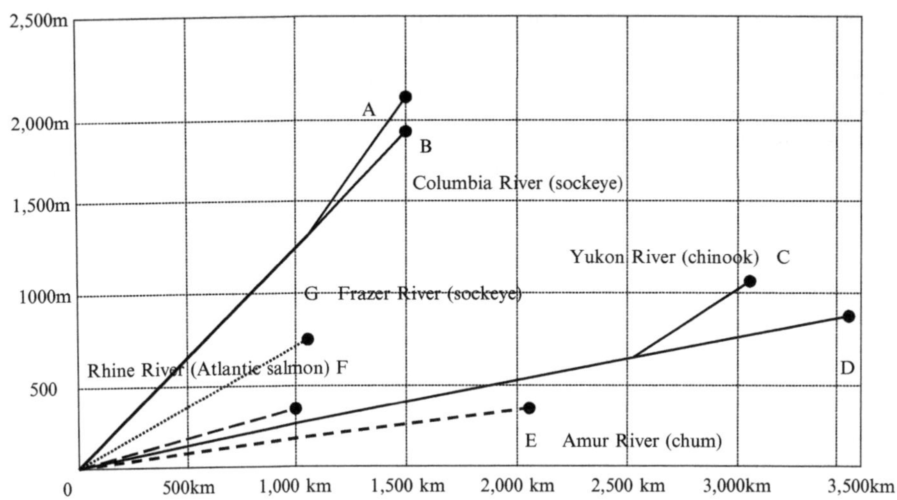

Fig. 10.1 Maximum fresh water distance and altitude of spawning runs of salmon. Sources of information: This figure is a revised version of Figure 1 in Murota (2003, p. 24). (A) Devine (1992), (B) Gross et al. (1998), (C) Mathews (1986), (D) Jarman (1972), (E) Smirnov (1976), (F) Pearce (1993), and (G) Johnston et al. (1997). Altitude of each point was estimated by searching the contour in the geological map corresponding to the area of such point

mum distances. Their altitudes are around 1,000 m above the sea level. Historical maximum in terms of altitude was recorded for groups of sockeye salmon (*Oncoryncus nerka*) in the Columbia River. As the points A and B in Fig. 10.1 show, maximum altitudes of their spawning run were as high as around 2,000 m (Devine 1992; Gross et al. 1998; quoted in Murota 2003, p. 24). The Amur River, being some 4,416 km long, has spawning runs of chum and pink salmon (*Oncoryncus keta* and *Oncoryncus gorbuscha*). On this river, Smirnov (1976) writes; "the autumn form (of chum) may swim as far as 2,000 km." The location of this maximum distance is the area of confluence of the Silka and Arguni Rivers, the place in East Siberia beyond Far East (Khantashkeeva and Murota 2004, pp. 56–57). Point F shows the historical case of the Rhine River which was a rich salmon river until a 100 years ago or so.

10.4 Quantitative Analysis of MDN Uploadings in the Russian Far East

In order to quantitatively understand the contribution of Pacific salmon to the MDN uploadings in the Russian Far East, we need the salmon escapement data for that area. The most recent data which the author could obtain thus far comes from the Pacific Research Fisheries Center (2003). Based on this data, the author derived Table 10.1, which shows escapements of adult Pacific salmon by species, regions, and sub-regions in 2002. Table 10.2 shows the ratio of escapements in the total spawning runs of salmon by species, regions, and sub-regions in the same year.

Table 10.1 Escapements of adult Pacific salmon to spawning areas in the Russian Far East by species, region and subregion in thousands of fish, 2002

Region, subregion	Pink	Chum	Sockeye	Coho	Chinook	Total
Western Bering Sea		740	195		935	
Eastern Kamchatka	2,993	1,006	385	31	64	4,482
Kurils	4,136	187				4,323
The Sea of Okhotsk						
Western Kamchatka	48,414	1,200	2,973	101	40	52,730
Continental coast	735	6,454		29		7,218
Sakhalin coast	8,077	113				8,191
Amur basin	1,632	1,339				2,972
The Sea of Japan						
Primorye	678	40				718
Southwestern Sakhalin	450					450
Total	67,117	11,082	3,554	161	105	82,021

Source: This table is the reproduction of Table 1 in Murota (2005) which was made from Table 8 in Pacific Research Fisheries Center (2003). In some places, the sum of the numbers in each column or row do not exactly coincide with the total number in the corresponding column or row because numbers less than 1,000 are omitted from listing in this table

Table 10.2 Ratio of escapements in the total spawning runs of salmon in the Russian Far East by species, region and subregion in thousands of fish, 2002, where the total spawning runs are defined as the sum of total catch and escapements (%)

Region, subregion	Pink	Chum	Sockeye	Coho	Chinook
Western Bering Sea		97.4	91.1		
Eastern Kamchatka	64.6	30.1	21.6	8.2	47.4
Kurils	13.6	9.8	0	0	
The Sea of Okhotsk					
Western Kamchatka	60.2	40.2	28.4	44.4	95.0
Continental coast	75.5	83.5	0	40.8	
Sakhalin coast	61.6	10.8			
Amur basin	55.7	71.6			
The Sea of Japan					
Primorye	22.5	0			
Southwestern Sakhalin					
Average of each species	47.7	51.2	28.4	23.6	58.6

Source: This table is the reproduction of Table 2 in Murota (2005) based on Table 1 and other data in Pacific Research Fisheries Center (2003)

Table 10.3 Marine-derived nitrogen (N) and phosphorus (P) uploaded onto Russian Far East lands by salmon escapements in 2002

Species	Escapements (thousand fish)	Average weight per fish (kg)	Total weight by species (ton)	N uploaded (ton)	P uploaded (ton)
Pink	67,117	1.47	98,661	2,296.3	354.1
Chum	11,082	3.57	39,562	1,201.4	142.0
Sockeye	3,554	2.78	9,800	297.6	35.1
Coho	161	3.28	528	16.0	1.8
Chinook	105	7.48	785	23.8	2.8
Masou	n.a.	2.50			
Far East total				4,535.1	535.8

Source: This table is the reproduction of Table 3 of Murota (2005) which was made from Tables 1 and 2 and other data explained in this paper. In some places, the sum of the numbers in each column or row do not exactly coincide with the total number in the corresponding column or row because numbers less than one ton are omitted from listing in this table

For preliminary quantification, the author focuses only on nitrogen (N) and phosphorus (P) among the many kinds of nutrients which are embodied in spawning salmon. Since there are no readily available data of N and P contents for adult Pacific salmon in Russian rivers, the author uses the data of Larkin and Slaney (1996) for Pacific salmon spawning in the rivers of British Columbia, Canada. Table 10.3 shows the result of these calculations.

Table 10.4 presents the author's previous results from the British Columbia case (Murota 2003, p. 21). Comparing Tables 10.3 and 10.4, we can tell that the contribution of Pacific salmon to the uploadings of MDN in the Russian Far East is about two times larger than that in British Columbia.

Table 10.4 Marine nitrogen (N) and phosphorus (P) uploading onto lands by five species of Pacific salmon in British Columbia (BC), Canada (estimates for the annual average for several years around 1990)

Species	Annual average escapements (thousand fish) (a)	Average body weight per fish(kg) (b)	Total weight by species (ton) (c)	Annual average N uploading (ton) (d)	Annual average P uploading (ton) (e)
Sockeye	10,000	2.27	22,700	689	81.5
Pink	20,000	1.82	36,400	1,105	130.6
Chum	1,500–2,500	5.45	8,200–13,600	249–413	29.4–48.5
Chinook	500	15.91	8,000	243	28.7
Coho	200–300	4.55	900–1,400	27–43	3.2–5.0
BC total				2,313–2,493	273–295

Remarks: This table is the reproduction of Table 2 in Murota (2003, p. 21). Data in (a) are taken from Henderson and Graham (1998). Data in (b) are taken from Larkin and Slaney (1996). (d) = (c) × 0.00359. These coefficients are adopted from Larkin and Slaney (1996). Though the numbers in columns (c), (d), and (e) have several digits for computation purposes, their significant digits are two or even one because the ones in column (a) are at most two

10.5 Material Cycles and Implications for a Sustainable Economy

In the studies of uo-tsuki-rin (fish breeding forest) in Japan, it has become evident that the supply of certain kinds of materials of terrestrial origin such as water-dissolved iron is essential for phytoplankton growth in the ocean. Poor phytoplankton growth in the ocean reduces fish stocks. To counter this problem, research efforts have been focused on the identification of terrestrial sources of iron supply to the ocean, especially to the Sea of Okhotsk. Wetlands, grasslands, and forests all appear to be important iron suppliers (Shiraiwa 2009 and others).

While this direction of research is very important, the research on material transport in the opposite direction is also important from the viewpoint of material cycling.

Thus far the amount of MDN transport by Pacific salmon has only been studied in the Russian Far East and British Columbia. Quantitative analyses are still needed for the U.S. states of Washington, Oregon, California, and Alaska. In Japan, the ratio of escapements in the total runs of salmon is very small compared with the high ratios in the Russian Far East, as shown in Table 10.2 of this paper, so that the amounts of MDN uploadings in Japan seem to be very small.

The contribution of Atlantic salmon has to be assessed as well for rivers which empty into the Atlantic Ocean. The St. Lawrence River in North America used to host a large number of spawning Atlantic salmon. There also are many salmon rivers in Europe. If we consider a single body of spawning salmon of a particular weight, the contribution of Pacific salmon to MDN transport onto land may be larger than that of Atlantic salmon because the former is semelparous while the latter is not.

After such a global scale analysis is completed, more detailed analysis of MDN transport by Pacific salmon in the Russian Far East should be conducted again to access the global and local significance of salmon rivers there from the viewpoint of the material cycle.

10.6 Conclusion

The number of naturally spawning salmon, both Pacific and Atlantic, has significantly decreased in the last 100 years or so in the northern hemisphere due to dam construction, water pollution, riverbed alterations of various forms and other anthropogenic changes. It is necessary for scientists, policy makers and the general public to review the full impacts of these alterations to naturally spawning salmon, as the role of the cross-boundary MDN transport to upstream areas by naturally spawning salmon is widely under-recognized. Environmental governance strategies to promote natural spawning will promote material cycling and support biodiversity, ecosystem health and the pursuit of a sustainable economy.

Acknowledgement The author is very thankful to Dr. Takayuki Shiraiwa of RIHN who gave him an opportunity to present an earlier version of this paper at the RIHN International Conference. This study partly benefited from the special research grant of 2010 for the promotion of the advancement of education and research in graduate schools provided by the Promotion and Mutual Aid Corporation for Private Schools of Japan and Japan's Ministry of Education, Culture, Sports, Science and Technology of Japan. Great appreciation is also due to an anonymous reviewer for valuable comments toward its improvement.

References

Bilby RE, Fransen BR, Bisson PA (1996) Incorporation of nitrogen and carbon from spawning coho salmon into the trophic system of small streams: evidence from stable isotopes. Can J Fish Aquat Sci 53:164–173

Bilby RE, Beach EW, Fransen BR, Walter JK (2003) Transfer of nutrients from spawning salmon to riparian vegetation in western Washington. Trans Am Fish Soc 132:733–745

Brock CS, Leavitt PR, Schindler DE, Quay PD (2007) Variable effects of marine-derived nutrients on algal production in salmon nursery lakes of Alaska during the past 300 years. Limnology and Oceanology 54(2):1588–1598

Cederholm CJ, Houston DB, Cole DL, Scarlett WJ (1989) Fate of coho salmon (*Oncoryncus kisutch*) carcasses in spawning streams. Can J Fish Aquat Sci 46:1347–1355

Cederholm CJ, Kunze MD, Murota T, Sibatani A (1999) Essential contributions of nutrients and energy for aquatic and terrestrial ecosystems. Fisheries 24(10):6–15

Devine B (1992) Salmon dammed. Audubon 94(1):82–89

Gross HP, Wurtbaugh WA, Luecke C (1998) The role of anadromous sockeye salmon in the nutrient loading and productivity of Redfish Lake, Idaho. Trans Am Fish Soc 127:1–18

Helfield JM, Naiman RJ (2001) Effects of salmon-derived nitrogen on riparian forest growth and implications for stream productivity. Ecology 82:2403–2409

Henderson MA, Graham CC (1998) History and current status of Pacific salmon in British Columbia. North Pacific Anadromous Fish Commission Bull. 1:13–22

Hilderbrand GV, Farley SD, Robbins CT, Hanley TA, Titus K, Servheen C (1996) Use of stable isotopes to determine diets of living and extinct bears. Can J Zool 74:2080–2088

Jarman C (1972) Atlas of animal migration. William Heinemann, London

Johnston NT, Macdonald JM, Hall KJ, Tschaplinski PJ (1997) A preliminary study of the role of sockeye salmon (*Oncorhyncus nerka*) carcasses as carbon and nitrogen sources for benthic insects and fishes in the 'Early Stuart' stock spawning streams, 1050 km from the ocean. Fisheries Project Report No. RD55. Fisheries Branch, Ministry of Environment, Lands and Parks, Province of British Columbia, Canada

Juday C, Rich WH, Kemmerer GI, Mann A (1932) Limnological studies of Karluk Lake, Alaska, 1926–1930. Bulletin of the Bureau of Fisheries (Bureau of Fisheries, US Department of Commerce) 12:407–436

Khantashkeeva T, Murota T (2004) Post-Soviet transition of the Russian Far East: its economic/environmental diversity in international perspectives. World Wide Bus Rev 5:1–127

Krokhin EM (1957) (English translation 1959) Sources of enrichment of spawning lakes in biogenic elements. Fisheries Research Board of Canada, Translation Series No. 207 (trans: Foerster RE)

Krokhin EM (1967) (English translation 1968) Effect of size of escapement of sockeye salmon spawners on the phosphate content of a nursery lake. Fisheries Research Board of Canada, Translation Series No. 1186. (trans: Foerster RE)

Larkin GA, Slaney PA (1996) Trends in marine-derived nutrient sources to south coastal British Columbia streams: impending implications to salmonid production, Watershed Restoration Management Report No. 3. Ministry of Environment, Lands and Parks and Ministry of Forests, British Columbia, Canada

Lessard JL, Merritt RW (2009) Investigating the effect of marine-derived nutrients from spawning salmon on macroinvertebrate secondary production in southeast Alaskan streams. J N Am Benthol Soc 28(3):683–693

Lessard JL, Mettitt RW, Cummins KW (2003) Spring growth of caddisflies (Limnephilidae: Trichoptera) in response to marine-derived nutrients and food type in southeast Alaskan stream. Annales de limnologie 52(4):1588–1598

Mathews R (1986) The Yukon. Holt, Reinhart, and Winston, New York

Mertz JE, Moyle PB (2006) Salmon, wildlife, and wine: marine-derived nutrients in human-dominated ecosystems of Central California. Ecol Appl 16(3):999–1009

Murota T (1998) Material cycle and sustainable economy. In: Keil R, Bell DVJ, Penz P, Fawcett L (eds) Political ecology: global and local. Routledge, London, pp 120–138

Murota T (2003) Marine nutrient shadow: global comparison of anadromous salmon, fishery, and guano occurrence. In: Stockner JG (ed) Nutrients in Salmonid ecosystems: sustaining production and biodiversity, American fisheries society symposium 34. American Fisheries Society, Bethesda, pp 17–31

Murota T (2005) Interactions between anadromous salmon and human beings from a viewpoint of nutrients cycle and environmental change. In: Proceedings of Russian National Workshop on Research related to the international human dimensions programme on global environmental change, Moscow, pp 292–299 (Russian translation, pp 143–152)

Nagasaka A, Nagasaka Y, Ito K, Mano T, Yamanaka M, Katayama A, Sato Y, Grankin AL, Zdorikov AI, Boronov GA (2006) Contributions of salmon-derived nitrogen to riparian vegetation in the northwest Pacific region. J Forest Res 11(5):377–382

Pacific Research Fisheries Center (2003). Biostatistical information on salmon catches, escapements, outmigrants number, and enhancement production in Russia 2002, NPAFC, Doc. 736, Vladivostok

Pearce F (1993) Greenprint for rescuing the Rhine. New Sci 138:25–29

Scheuerell MD, Levin PS, Zabel RW, Williams JG, Sanderson BL (2005) A new perspective on the importance of marine-derived nutrients to threatened stocks of Pacific salmon (*Oncorhyncus* spp.). Can J Fish Aquat Sci 62(5):961–964

Schindler DE, Leavitt PR, Brock CS, Johnson SP, Quay PD (2005) Marine-derived nutrients, commercial fisheries, and production of salmon and lake algae in Alaska. Ecology 86(12):3225–3231

Shiraiwa T (2009) Giant fish breeding forest, a paper read at IGU/LUCC International Conference organized by Pacific Institute of Geography. Far Eastern Branch of Russian Academy of Sciences, Vladivostok

Smirnov AI (1976) Chum salmon, Great Soviet Encyclopedia, 7th edn. Macmillan, New York, 87p

Uchiyama T, Finney BP, Adkison MD (2008) Effects of marine-derived nutrients on population dynamics of sockeye salmon (*Oncorhyncus nerka*). Can J Fish Aquat Sci 65(8):1635–1648

Zhang Y, Negishi JN, Richardson JS, Kolodziejczyk R (2003) Impacts of marine-derived nutrients on stream ecosystem functioning. Proc Biol Sci 270:2117–2123

Part IV
Impacts of Human-Made Boundaries

Chapter 11
Introduction

Stephen McCauley

Abstract This chapter synthesizes important themes raised in the chapters on the dilemmas of human-made boundaries in water management. A fundamental dilemma in water management is that human-made boundaries – either legal-political boundaries or management boundaries – fail to correspond to the complex and continuous ecological systems that are meant to represent. This mismatch results in negative externalities and inefficient resource use. The chapters in this section show how these inefficiencies emerge in different water management contexts, and they all point to the need for new governance models that address these boundary issues in water management.

Keywords Human-made boundaries • Externalities

11.1 Overview of Human-Made Boundary Cases

The chapters in this section reveal the ways in which the human tendency to divide complex ecological systems—to assert ownership or to manage—partitions whole ecosystems in ways that conflict with ecological processes and lead to sub-optimum resource management outcomes. As Prof. Endo Takahiro says in Chap. 15, "Human-made boundaries often become obstacles to effective water management."

The chapters highlight two cross-boundary issues that often go unseen in environmental management: the linkage between land-use and ocean productivity and linkages involving groundwater systems. The Amur-Othotsk Project, described by Prof. Hanamatsu in Chap. 12, examines the human boundary issues at play in the "Giant fish-breeding forest," the massive ecological system linking land-use and land cover

S. McCauley (✉)
George P. Marsh Institute, Clark University, 950 Main Street, Worcester, MA 01610, USA
e-mail: smccauley@clarku.edu

M. Taniguchi and T. Shiraiwa (eds.), *The Dilemma of Boundaries: Toward a New Concept of Catchment*, Global Environmental Studies,
DOI 10.1007/978-4-431-54035-9_11, © Springer Japan 2012

changes throughout large parts of China, Russia, and Mongolia, as well as economic activities in Japan with biomass production in the Othotsk Sea. In Chap. 13, Prof. Wakana provides historical context for this land–sea linkage, detailing the historic development of the fish-breeding forest (Uosukirin) management system in Japan. In Chap. 14, Prof. Jarvis examines the dynamic spatial, temporal and political dimensions of groundwater boundaries, and in Chap. 15, Prof. Endo describes the case of Saijo City, Japan, where a legal boundary between surface water and groundwater users results in inefficient use of the integrated water system.

The chapters reveal the core problems associated with human-made boundaries in ecosystem management: the lack of correspondence between politico-jurisdictional boundaries and ecological systems, and the problem of segmented or fragmented boundaries within management regimes. The resulting management problems are the inefficiencies associated with externalities. Despite these difficulties, the chapters all offer guidance for moving forward, toward integrated management approaches that blur these artificial boundaries and internalize the externalities.

11.2 Conflicts Arising from Legal–Political Boundaries

Political jurisdictions have historically been created using waterways as boundaries. This practice has resulted in a legacy of spatially segmented decision-making in which the interests of individual jurisdictions are at odds with the interests of the watershed system as a whole. In the case of the "Giant Fish-Breeding Forest," (GFBF) sustained ocean productivity is critical to all of the countries involved, yet the different locations of the countries within this huge ecological system—either in the upper basin or in the sea—mean there is a discord between benefits and costs, which leads to free rider tendencies. The case of the water dispute in Saijo City, Japan also reveals how jurisdictional and legal boundaries that are at odds with ecosystem processes result in negative externalities. Plans by Matsuyama City, Saijo City's neighbor to the west, to ensure supply by increasing diversion of water, conflict with Saijo City's reliance on the river for groundwater recharge. The problem, which is exacerbated by an institutional arrangement that assigns public water rights to surface water users but not groundwater users, is rooted in the historical practice of using waterways as political boundaries between jurisdictions. Furthermore, effective environmental monitoring and cooperation across political boundaries is difficult and, as the GFBF case shows, these challenges are exacerbated when border areas are in dispute or when relations are strained.

11.3 Conflicts Arising from Management Boundaries

While legal–political boundaries—either at international or intra-national levels—present obvious challenges, pervasive conflicts also arise from the boundaries designed into environmental management systems. Chapter 14 describes this discord between ecosystems and management systems as one between bona-fide boundaries (those with a physical basis such as recharge areas, hydrogeologic regions defined by storage characteristics, watercourses or catchments) and fiat boundaries (user-created boundaries such as the artificial discharge areas created by wells, drainage areas created by large resource development projects, and wellhead or other conservation areas). Management system boundaries conflict with ecosystem processes either due to ineffectual design, imperfect understanding of system dynamics, or the fundamental difficulty of designing management systems that capture the continuity of physical processes.

Chapter 12 describes the problem of sectionalism, or fragmentation, in which management systems are in place but are either incomplete, competing, or overlapping. Numerous resource management regimes have been implemented in the Amur River basin over recent decades, including multi-national agreements like the Ramsar Convention on wetlands, bilateral agreements providing for environmental cooperation among neighboring countries, and domestic forest and wetlands management policies. While offering some protection, this system of incremental, fragmented management is inadequate for managing the huge and complex GBFB ecosystem as a whole. At its worst, it enables sectionalism, in which multiple agencies within the same country have authority over a resource, often with conflicting goals.

The two chapters on groundwater highlight the failure of management systems to deal with the spatial dimensions of three dimensional resources. Though groundwater and surface water are part of a unified system, in the Saijo City case the legal status of water depends on whether water is on the surface or below ground. In this case, the drawing of a vertical boundary results in differing legal statutes for surface water and groundwater. On the other hand, Prof. Jarvis offers several examples of groundwater domains that vary over the vertical dimension but for which management boundaries are only horizontal, and he proposes a more three dimensional view that distinguishes between shallow and deep aquifers.

Management inefficiencies also result from incomplete knowledge of ecosystem dynamics. The fragmented ecosystem management regimes that currently exist in the Amur River basin were developed without understanding of such continental scale processes as the role of dissolved iron transport in linking land-use and ocean productivity. The Amur-Othotsk project now provides an important scientific case for re-thinking ecosystem management at the continental level. Similarly, Prof. Wakana's description of the development of the original fish-breeding forest (Uosukirin) system in Japan shows how implementation of this management system has increased over the centuries as the linkage between upland forests and fisheries has been better understood. In the Saijo City case, the legal construct that separated surface water user rights and groundwater user rights was based on an incomplete understanding of the role of surface water in groundwater recharge. From a conceptual perspective, Chap. 14 argues that resource management regimes often fail to appreciate the ways in which political and user domains significantly alter the nature of groundwater systems.

11.4 The Problem: Externalities

The central problem arising from these cross-boundary management conflicts is that they produce negative externalities which result in sub-optimum resource use. Where users are able to utilize resources without paying the full costs associated with their use, the resource is used to excess and overall welfare worsens. In the case of the Amur-Othotsk Project, land-use among countries in the Amur River basin affects the Sea of Othotsk fisheries on which the four nations depend economically. The externality in the Saijo City case results from the fact that the marginal costs to surface water users in Mayamota City do not consider the costs to Saijo City groundwater users and therefore result in inefficient use of the overall water system. In revealing these instances of negative externalities, the chapters point to an important goal of ecosystem management: to design systems which, by better integrating natural and human-made boundaries, internalize such externalities.

11.5 Toward Improved Human-Made Boundaries

The insights generated from these chapters suggest the need for new governance models, and each chapter proposes management approaches that may address some of the inefficiencies associated with human-made boundaries. Prof. Wakana's chapter on the Uosukirin in Japan describes the revival of the traditional management system linking forests and fisheries and the more recent expansion of the system from waterfront areas into larger woodland areas. Based on the finding that land–ocean linkages operate at a continental scale through the transport of dissolved iron, the Amur-Othotsk Project is spear-heading a consortium among research institutes, universities, governmental organizations, international organizations and private companies—the Giant Fish-Breeding Forest Partnership—which aims to establish an intergovernmental regional framework for information exchange and cooperation between the countries.

The chapters on groundwater systems also provide guidance for new management systems. Prof. Jarvis proposes a management paradigm based on blurring the boundaries of resource domains and user domains to consider them as problemsheds rather than watersheds, and he describes some management approaches that do blur this boundary, such as unitization and hydrogeologic reserves. Prof. Endo sees potential in the now well-established integrated water resource management (IWRM) approach, though he emphasizes that the necessity of IWRM lies not only in the physical unity of groundwater and surface waters, but in the need to design social institutions that internalize externalities and provide for social welfare.

The proposed solutions tend to embrace integrated water resource management principles but to expand their scope to include the lessons derived from these cross-boundary studies: the large scale of the ocean–land linkage, the critical role played by fisherman in the Uotsukirin, the shifting boundaries of user domains and groundwater resource domains, and the role of legal boundaries in shaping the efficiency of water systems. These insights point to a final lesson expressed in these chapters: the role of scientific understanding in designing effective resource management regimes. Better understanding of the linkages across human and ecological boundaries allows for management regimes that better internalize the externalities that often plague resource management.

Chapter 12
National Boundaries and the Fragmentation of Governance Systems: Amur-Okhotsk Ecosystem from the Legal and Political Perspective

Yasunori Hanamatsu

Abstract The Amur-Okhotsk Project (AOP) has recently found that primary production in the Sea of Okhotsk and the neighboring Oyashio region has been dependent on dissolved iron transport from the Amur River. In the Amur River basin, dissolved iron originates mainly from wetlands and forests. Our project has clarified this ecological linkage between the continent and open waters, and proposed a new concept, the *"Giant Fish-Breeding Forest" (GFBF)* system. The implication is that in order to conserve the marine resources in the Sea of Okhotsk and the Oyashio region, it is also necessary to protect the environment in the Amur River basin, especially wetlands and forests.

From the legal and political perspective, this "Giant Fish-Breeding Forest" system has two different kinds of boundaries inside itself; national boundaries (transboundary) and regime boundaries (fragmentation of governance systems). The conservation of this ecosystem has been disrupted as a result of these human-made boundaries. This ecosystem spans borders of China, Russia, Mongolia and Japan, but these countries do not share the same benefits or costs of conserving this system, leading to "free riding." In addition, some aspects of the environment in this region have already been regulated by international and national laws, but these management regimes have either concluded or have been implemented independently, with some overlap and conflict, and therefore they are inadequate for the conservation of the whole system of the "Giant Fish-Breeding Forest" as a single unit.

Discussing the problems resulting from these two human-made boundaries, this paper demonstrates that, in order to manage this ecosystem effectively, it is important to coordinate existing legal systems and policies in an integrated manner and to promote a common understanding among countries in this ecosystem.

Y. Hanamatsu (✉)
Slavic Research Center, Hokkaido University,
Kita-9, Nishi-7, Kita-ku, Sapporo 060-0809, Japan
e-mail: hanamatsu@slav.hokudai.ac.jp

M. Taniguchi and T. Shiraiwa (eds.), *The Dilemma of Boundaries: Toward a New Concept of Catchment*, Global Environmental Studies,
DOI 10.1007/978-4-431-54035-9_12,

Keywords "Giant Fish-Breeding Forest" • National boundaries • Regime boundaries • Fragmentation • Coordination • Integrated management

12.1 Introduction

Ecological systems based on natural processes such as water circulation fundamentally recognize no boundaries, but from the earliest human history, human beings have drawn various socio-political boundaries between, for example, themselves, communities, territories, or cultures (Kolossov 2005). Due to these boundaries, the conservation of ecological systems as whole units has become a difficult task. The area of focus of this paper, the Amur-Okhotsk ecosystem, is a typical example. The Amur-Okhotsk Project (AOP) has recently found that primary production in the Sea of Okhotsk and the neighboring Oyashio region has been dependent on dissolved iron transport from the Amur River basin (Fig. 12.1).

The Sea of Okhotsk and the Oyashio region are known to be among the most productive oceanic areas in the world. Almost all of these areas are within the Russian exclusive economic zone (EEZ), and include the most important fishing areas, with almost 60% of the Russian national catch. Approximately 50% of the sea product of Japan is from these areas. A share of the processed seafood from Russian-origin raw material finds its way back to the booming Russian market. China, for instance, is the key supplier of fillets of Alaska pollock to Russia with

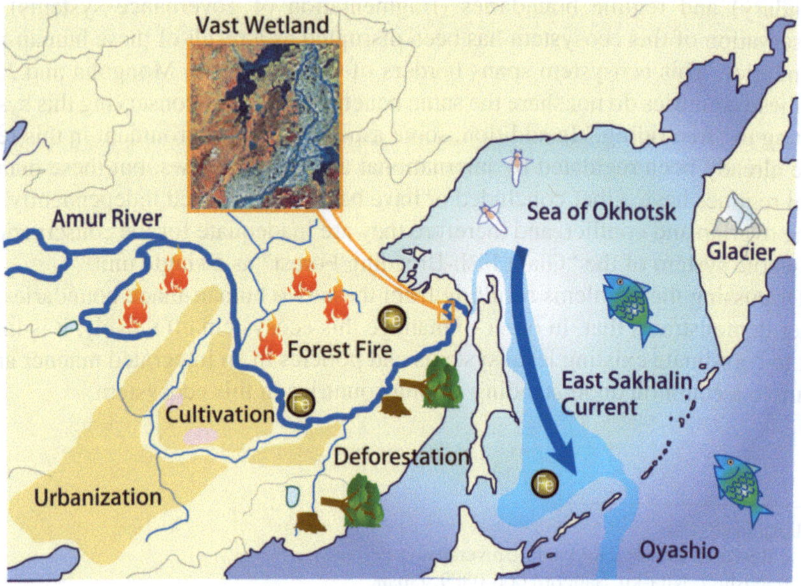

Fig. 12.1 Amur-Okhotsk ecosystem

close to 15,000 tonnes of exports in 2006. In short, the national economy and food security of these countries are highly dependent on fishing and fish products in these oceanic areas (FAO/EBRD 2008).

While the oceanic areas are important economically to several countries in the area, in the Amur River basin dissolved iron originates from the wetlands and forests, located mainly in the Russian Far East and the Northeast China. It has been pointed out that the recent excessive land cover and land-use changes in the Amur River basin may affect the flux of dissolved iron and may result in biomass production changes in the Sea of Okhotsk and Oyashio region. The Amur River basin is currently disturbed by various human and natural impacts such as forest fires, deforestation, illegal logging, land-use change from natural wetlands to agricultural fields, other agricultural and industrial activities, flooding and drought. Therefore, in order to conserve the sustainability of the marine ecosystem in the Sea of Okhotsk and the Oyashio region, the negative impact of land-use changes in the inland areas of this region, as well as over-fishing in the sea of Okhotsk and the Oyashio region, must be mitigated (Shiraiwa 2006; Onishi 2007).

The AOP has clarified this huge ecological linkage between the continent and open waters, and proposed a new environmental concept, the *"Giant Fish-Breeding Forest"* (GFBF) system, by expanding the traditional Japanese idea of *Uotsuki-Rin* (fish-breeding forests), which has related upstream forests with the coastal ecosystem both physically and conceptually. What is needed next is a discussion of a possible conservation strategy for the protection of the whole ecosystem of the GFBF.

Conservation of this ecosystem has been disrupted as a result of two different human-made boundaries. These boundaries, *national boundaries (transboundary)* and *regime boundaries (fragmentation of governance systems)*, each have significant implications. This paper focuses on the influence of these two different kinds of human-made boundaries on the conservation of this ecosystem. While discussing the problems resulting from these two human-made boundaries, this paper illustrates that, in order to manage this ecosystem effectively, it is important to coordinate existing legal systems and policies in an integrated manner and to facilitate a common understanding among countries in this ecosystem.

12.2 National Boundaries Between Four Nations in the GFBF Ecosystem

12.2.1 Transnational Nature of the GFBF Ecosystem

The GFBF and its impacted area encompass nearly four million km^2. It includes parts of the territories of Mongolia, China and Russia as well as the Russian and Japanese exclusive economic zones (EEZ). In other words, these areas comprise a huge ecological system that spans the borders of China, Russia, Mongolia and

Japan. However, these countries do not share the same benefits or costs of conserving this ecosystem.

The primary beneficiary state is considered to be Russia, whose catch is highly dependent on the fish resources in the Sea of Okhotsk. The fishing industry is crucial not only for the food supply for the Russian people, but also for the national economy since approximately half of the total Russian catch is exported from these particular oceanic areas (FAO/EBRD 2008; Oude Elferink 2001). However, this does not mean that other states do not have a claim to share the benefits of fish products in these areas. China and Japan have also enjoyed the benefits of these fish resources, but less so than Russia. A large amount of fish have been exported from the Russian Far East to these two countries.

The states that have contributed most to the transportation of the dissolved iron from the Amur River basin to the ocean and which also may negatively impact this transport are Russia and China. The main sources of dissolved iron are wetlands and forests in the Russian Far East and Northeast China. The land-use change from wetlands to agricultural field and excessive logging and deforestation in these areas may threaten the ecological system of the GFBF. From this perspective, Russia and China should make efforts to minimize effects on this ecosystem.[1]

At first glance, Japan may appear to be only a beneficiary of the fish resources in the Sea of Okhotsk and the Oyashio region. It is clear that Japan does not supply dissolved iron to these oceanic areas at such a large scale as the Amur River basin. However, it has been recently indicated that Japan has given financial and technical assistance to Chinese development programs, including farmland development activities in the Sanjiang Plain wetlands. In addition, Japan has imported vast amount of agricultural products from Northeast China, and timbers from the Russian Far East. Therefore, Japan may have a serious effect on the dissolved iron transport in the Amur River basin. It can be said that Japan is also one of the key players in this ecosystem.

In this situation, the conservation of the whole system of the GFBF cannot be achieved without mutual cooperation among the four countries. In international relations, every country usually wants to be a free-rider for the use of shared natural resources. For example, even if Russia implements strict regulations for the sustainable management of forests and fish resources, China still may promote wetland development policies resulting in a decrease of the dissolved iron concentration in the future. In this case, until the fish stock is exhausted, China can continue to enjoy these fish resources without contributing to its conservation. This is the so-called free-riding problem (Kaul et al. 1999). When faced with the possibility of China's free-riding, Russia may opt to stop its conservative policy and place a fish export ban on China. In this sense, the GFBF ecosystem cannot or will not be adequately addressed by individual countries acting alone. This means that conserving

[1] It is conceivable that Mongolia also has some influence on the flux of the dissolved iron in the Amur River basin. However, the impact on the GFBF system has not yet been assessed in the AOP, and it is estimated that the impact is limited.

this system requires international or regional cooperation among these four nations. Unfortunately, in spite of the importance and uniqueness of this ecosystem, there has been no cooperative framework for the conservation of this ecosystem among these countries.

12.2.2 The Amur River Basin

Moreover, the Amur River has been under extreme political tension since the middle of the nineteenth century and there has been little transboundary cooperation. This situation has resulted in the Amur River becoming one of the most seriously polluted waters in Russia (Fig. 12.2).

The Amur River has been the site of political boundaries between China and Russia since the Treaty of Aigun and the Convention of Peking signed by the two countries in 1858 and 1860, respectively. Since the two agreements were considered to be unequal treaties, the boundary was rather unstable until the two countries finally agreed to define the boundary in 2004 (Iwashita 2005). This history has made the Amur River one of the most difficult rivers to monitor for conservation purposes.

At present, the upper and middle reaches of the Amur River mark the border between China and Russia as well as the border between the European and Asian regions. There are many cases in the world where transboundary resources such as rivers and lakes are used as natural fences for designating boundaries between neighboring countries (Benvenisti 2002; Querol 2005). In many cases, these boundary resources have become detrimental to cooperation.

The demarcation dispute in the Amur River basin began to improve in the 1990s as a result of a political compromise between China and Russia. The two countries ratified the Treaty on Good Neighbor Relations, Friendship and Cooperation in February 2002, which includes cooperation on transboundary water protection. The two governments also appointed respective official departments for monitoring the Amur River and the Ussuri River. The concerned departments in China and Russia have monitored the waters of these rivers eight times so far.

In November 2005, a notorious explosion accident occurred involving a petrochemical company in Jilin Province of China. This accident significantly polluted the Songhua River, the largest tributary of the Amur River (UNEP 2006a). The Chinese State Environmental Protection Adiministration (SEPA) invited an expert team from the United Nations Environmental Programme (UNEP) to conduct a field evaluation of the affected region. Cooperation between the Chinese and Russian governments on transboundary water protection also contributed to solving this problem (Zaisheng et al. 2008). China and Russia agreed to set up a joint monitoring team and implemented the monitoring measures beginning in December 2005.

The accident motivated both countries to collaborate to prevent transboundary pollution of the Amur River. In 2007, the two countries began annual joint-monitoring of the water quality in the Amur River. This case shows that detrimental

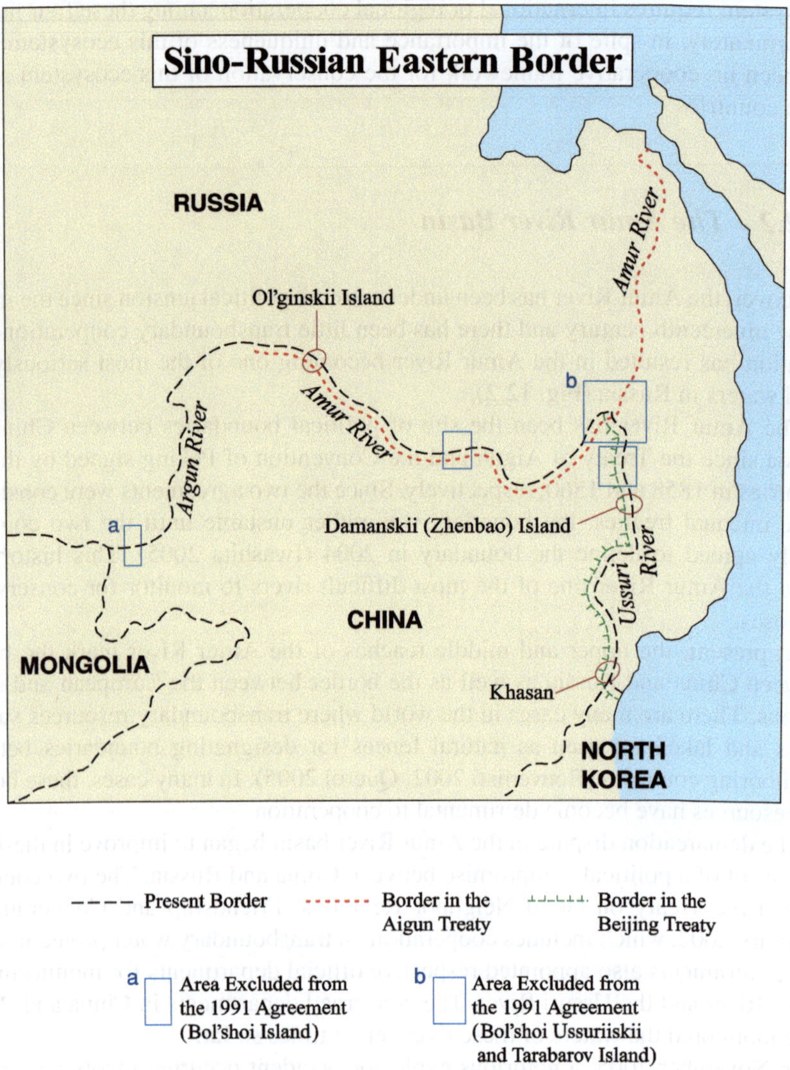

Fig. 12.2 Sino-Russian border and the Amur river (from Iwashita 2005, p. 96, with permission)

accidents can serve as an effective incentive for mutual cooperation and conservative measures. Other foreign countries and scientists were still not invited to join this program, but the accident contributed to the further conservation of the Amur River by increasing people's consciousness of the shared environment (Faure and Ying 2008).

Except for water pollution, no other threat to the GFBF ecosystem has yet stimulated conservation concern. Furthermore, there has been no cooperative framework for the conservation of this ecosystem among these countries. In this situation, more data and information is needed in order to clarify the extent to which land surface disturbances such as land-use change from wetlands to farmlands, forest fire and logging could adversely affect primary production in the Sea of Okhotsk and the Oyashio region directly or indirectly. For this purpose, joint-monitoring and research in the Amur River and its basin should be undertaken.

It remains difficult to do such work particularly in political boundary areas such as the Amur River because the resolution of the territorial dispute does not always promote an overall friendly cooperation for environmental protection measures in such areas. Therefore, new cooperative framework should be established for further research and joint-monitoring concerning the transportation of the dissolved iron, based on the shared understanding that their failure to take measures for the conservation of the GFBF ecosystem may have a negative effect on both countries in the future.

12.2.3 The Sea of Okhotsk

While the Amur River has been a source of political dispute between Russia and China, the Sea of Okhotsk has been a political hot spot between Russia and Japan. Russia has jurisdiction with regard to marine scientific research and the protection and preservation of the marine environment in the Sea of Okhotsk. In performing its jurisdiction, Russia is obliged to show due regard for the rights and duties of other states.[2] Owing to persistent territorial conflict, however, it was practically impossible for Japan to join the monitoring of environmental problems in the Sea of Okhotsk during the last half of the twentieth century. A collaborative effort between Russia and Japan for the environmental conservation of the Sea of Okhotsk is urgently needed to address the increasing activities relating to oil mining and natural gas exploration in the Sea of Okhotsk and its vicinity (UNEP 2006b) (Fig. 12.3).

Recently, Japan and Russia signed onto a cooperative program on research, conservation and sustainable use of the ecosystems in the adjacent areas.[3] This includes cooperation concerning joint research, monitoring and information exchange on the ecosystem in the Sea of Okhotsk and the Oyashio region. The program specifically refers to research concerning the impact of the Amur River basin upon the Sea of Okhotsk. While this program is neither legally binding nor constitutes an international agreement, it is expected that the program could promote a cooperative framework for environmental protection in these oceanic areas and enhance mutual understanding of the GFBF ecosystem.

---

[2] United Nations Convention on the Law of the Sea (UNCLOS,1982), Article 56 (1)(b) and (2).

[3] *Asahi News Paper*, 13 May 2009. The document has not yet been published (copy on file with author).

Fig. 12.3 The Map of the Sea of Okhotsk and the Russian jurisdiction

These cooperative frameworks have just started recently. Needless to say, such efforts are not sufficient in order to examine the exact ecological conditions of the ecosystem. Therefore, we must further our efforts to develop cooperative frameworks and measures.

In short, national boundaries within the GFBF system have long been an obstacle for the conservation of this system. No single country can cope with this entire ecosystem. Without mutual cooperation, free-riding problems may occur. In addition, national border conflicts have made international environmental cooperation in the boundary areas more difficult. Also, it is quite difficult to conduct joint-monitoring and research activities in the border areas. Therefore, what is needed is to "blur" the strength of boundaries by establishing an international cooperative framework. Atleast, joint-monitoring, data exchange and mutual communication will be indispensable as a starting point for the protection of shared ecosystems. Such efforts will promote a common understanding as to the importance of conserving this ecosystem among countries.

12.3 Fragmentation of Environmental Governance Regimes in the GFBF System

As explained in the previous section, the GFBF ecosystem is the enormous ecological system that cuts across the national boundaries among Russia, China, Mongolia and Japan. There is another set of boundaries that has disrupted the effective conservation of the GFBF ecosystem; boundaries between environmental protection rules, regimes and authorities.

Parts of the environment in the GFBF system have already been regulated by international and national laws and policies (Fig. 12.4), but these management regimes have been developed and implemented independently, sometimes with overlap or conflict, and therefore are not adequate for the conservation of the GFBF as a whole system. In recent years, this phenomenon has been observed in other environmental management cases or other issue areas, and is generally described as "fragmentation", "sectionalism" or "inter-linkage" problems in legal and political discourses (Wolfrum and Matz 2003; Chambers 2008; Oberthür and Gehring 2006). Nevertheless, little attention has been paid to this issue concerning environmental protection in the GFBF ecosystem.

When analyzing the existing international and domestic laws or policies that seem to be applicable for the conservation of this system, the question that must be addressed is whether existing legal systems and policies are adequate for the conservation of the whole system of the GFBF. According to the scope of application, there are three possible categories of laws or policies; global multilateral environmental treaties, bilateral environmental agreements, and domestic laws and policies in each country. The analysis of each category will help us understand what is needed for an effective management of this ecological system as a whole.

12.3.1 Global Multilateral Environmental Treaties

International law has traditionally regulated the use of natural resources indirectly by determining the basis on which rights are allocated among states. The legal status of natural resources varies according to whether the resource is under the sovereignty of one state, shared by several states, or held in common for the benefit of all. In general, it was assumed in the early development of international law that the control of natural resources is within sovereignty over land territory and territorial seas (Brownlie 1979). This tradition was reinforced by two famous UN General Assembly Resolutions: Permanent Sovereignty over Natural Resources[4] and Declaration on the Establishment of a New International Economic Order (NIEO).[5]

[4] UN General Assembly Resolution 1803 XVII (1962).

[5] UN General Assembly Resolution 3201 S-VI (1974).

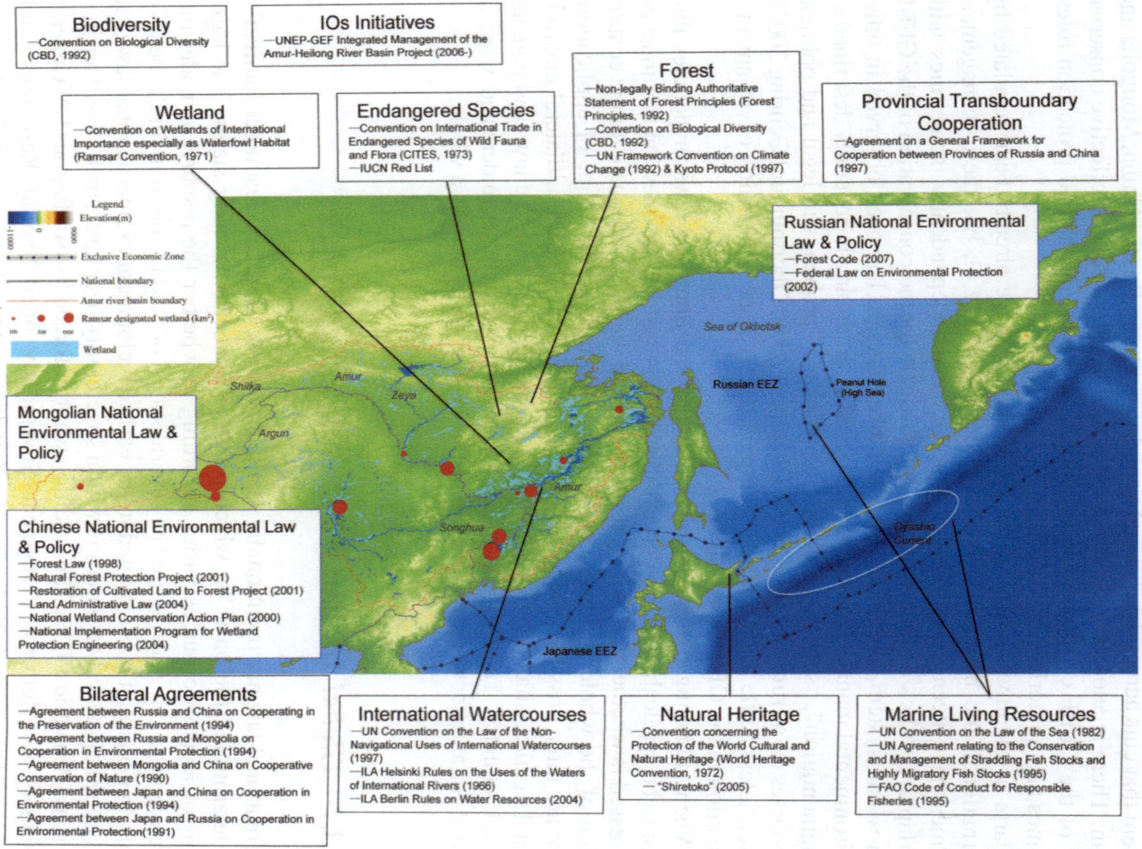

Fig. 12.4 Existing environmental regimes in the GFBF system

Therefore, in principle, states have the freedom to exploit resources within their territories and territorial seas, unless their exploitation could harm other states. However, recent environmental concerns have eroded state sovereignty and involved a redefinition of sovereignty itself. In the contemporary situation, states only have limited sovereignty over resources within their jurisdictions in so far as their use of the resources is against international rules and principles applicable to them (Schrijver 1997).

Global multilateral environmental treaties applicable to the GFBF system include the law of international watercourses, the law of the conservation of nature, species and biodiversity, and the law of living marine resources.

First, the major instruments of the law of international watercourses are the United Nations (UN) Convention on the Law of the Non-Navigational Uses of International Watercourses (1997) and the United Nations Economic Commission for Europe (UNECE)'s Convention on the Protection and Use of Transboundary Waters and Lakes (1992). The former is not yet in force, and China voted against the adoption of this convention at the UN General Assembly (Benvenisti 2002). The latter has been drafted in the European context, and ratified by Russia, but is not applicable to China, Mongolia and Japan. Another instrument is the International Law Association (ILA)'s Helsinki Rules on the Uses of the Waters of International Rivers (1966). This instrument is non-binding in itself but is considered to be customary law which is generally recognized to be legally binding on all states.

These rules are based on the broader concept of the "drainage basin," which is a geographical area "determined by the watershed limits of the system of waters, including surface and underground waters, flowing into a common terminus."[6] The problem is, however, that all these instruments are exclusively concerned with allocating water supply between upstream and downstream states, or preventing pollution or damage (McCaffrey 2007; Ying 2008), and do not address such issues as the transportation of dissolved iron, much less the conservation of the ecosystem as a whole.

Second, the law of conservation of nature, species and biodiversity includes the Convention on Wetlands of International Importance especially as Waterfowl Habitat (Ramsar Convention 1971); the Convention concerning the Protection of the World Cultural and Natural Heritage (World Heritage Convention 1972); the Convention on International Trade in Endangered Species of Wild Fauna and Flora (CITES 1973); the Convention on Biological Diversity (CBD 1992), and the Non-legally Binding Authoritative Statement of Forest Principles (1992).

The Ramsar Convention aims to conserve wetlands and promote the wise use of wetlands in a sustainable way. More than 15 Ramsar sites, including a small part of the Sanjiang Plain in China, have already been listed for their international importance (Simonov and Dahmer 2008). However, the Ramsar Convention is inherently limited by its segmented approach to the landscape. The Ramsar Convention in

[6] ILA Helsinki Rules, Article 1 (1).

principle focuses narrowly on land and water use within wetlands rather than the broader catchments of which they are an intimate part. The Convention recently has adopted the integrated river basin management program which links wise use of wetlands with river basin management and protection of biodiversity (Simonov and Dahmer 2008). This program is particularly relevant for the conservation of the GFBF system in so far as it can link wise use of wetlands with river basin management. It is still uncertain, however, whether this basin management approach may expand its scope to include the relationship between wetlands and the ocean environment in view of the ecological linkage identified as the GFBF.

The Convention on Biological Diversity has through the implementation process taken the "ecosystem approach", which is, by definition, "a strategy for the integrated management of land, water and living resources that promotes conservation and sustainable use in an equitable way".[7] Nevertheless, it is not yet clear whether this approach serves as a practical guide for the conservation of a huge and complex system such as the GFBF.

The Forest Principles are of limited legal authority and content reflecting the absence of international consensus on the subject. The Principles do not "internationalize" forest issues, and instead note that "their sound management and conservation is of concern to the governments of the countries to which they belong".[8] Forest conservation is only indirectly regulated by the CBD as one of the components of biodiversity, or by the UN Framework Convention on Climate Change (UNFCCC 1992) and the subsequent Kyoto Protocol (1997) as one of the means for the mitigation of climate change.

Thirdly, the UN Convention on the Law of the Sea (UNCLOS 1982) and the subsequent UN Agreement relating to the Conservation and Management of Straddling Fish Stocks and Highly Migratory Fish Stocks (Straddling Stocks Agreement 1995) regulates the management of living marine resources in this area. Almost all of the Sea of Okhotsk and the Oyashio region are within the Russian exclusive economic zone (EEZ). Therefore, Russia has the primary obligation and jurisdiction for the conservation and optimal utilization of the living marine resources in this area according to its capacity to harvest. Although other states may in certain circumstances have a claim to share fishing in the Russian EEZ, in principle other states have neither freedom of fishing nor unfettered freedom to conduct scientific research in the area.

Concerning the linkage between land and ocean, these instruments only establish a general framework for the regulation of land-based sources of marine "pollution", and therefore do not refer to the relationship between the conservation of the marine resources and any land-based substances which sustain marine resources, such as dissolved iron, rather than cause "pollution" (Hassan 2006; Mensah 1999).

---

[7] CBD, COP 5 Decision V/6 (2000), A1.

[8] Forest Principle, Preamble para.(f).

12.3.2 Bilateral Environmental Agreements

A few bilateral environmental agreements among countries in Northeast Asia have been concluded since the early 1990s. These include agreements concerning general environmental cooperation, and monitoring of transboundary waters, fisheries and forests in the Amur River basin, including the Agreement on Cooperation in Protection of the Natural Environment between China and Russia (1994); the Agreement on Cooperative Conservation of Nature between China and Mongolia (1990); and the Agreement on Cooperation in Protection of the Natural Environment between Russia and Mongolia (1994). These agreements have been effectively implemented by the appointed departments of each country. In addition, the Agreement on a General Framework for Cooperation between Provinces of Russia and China (1997) permits provinces to develop any international agreements across the river that do not encroach on exclusive responsibilities of central governments, and to establish cooperative mechanisms to implement these agreements. Based on this framework, some provinces in Russia and China have concluded agreements concerning primarily the prevention of transboundary water pollution.

Japan has concluded bilateral environmental agreements with China (1994) and Russia (1991). They are, in principle, general agreements on environmental cooperation, while focusing on some specific issues such as air pollution, climate change, energy security and biodiversity, as well as water pollution. Moreover, since the adoption of these agreements, these countries have paid attention to other environmental issues such as technical cooperation on water quality monitoring in China and the water pollution problem in the Amur River.

While it should be said that these bilateral agreements are successful, they have been formulated and implemented without taking into account the ecological linkage between land and ocean in the GFBF ecosystem. Moreover, considering the fact that the GFBF ecosystem is shared by the four countries, taking measures only under these bilateral agreements is not sufficient for the effective monitoring and management of the ecosystem. As mentioned above, data sharing and joint monitoring among the four countries concerning the transportation of dissolved iron in the Amur River basin and the Sea of Okhotsk are crucial for promoting the conservation of the ecosystem. In this sense, we should establish another multilateral framework or agreement under which all relevant countries can cooperate together for promoting conservation measures beyond each bilateral agreement.

12.3.3 Domestic Laws and Policies

In addition to the international and regional levels, environmental protection policies have been taken at the national level as well. Domestic laws and policies relevant to this system concern particularly the conservation or sustainable use of wetlands and forests in Russia and China, because such policies in these two countries could have a significant impact on the sustainability of the GFBF system. Mongolia also shares the continental part of this ecosystem, but its negative impact on the system is now considered to be negligible because of its lower level of industrialization. Therefore, in this paper, attention is paid to Russian and Chinese legal systems and policies.

Russia It is commonly acknowledged that forest resources in Russia have been degraded with the major causes of degradation being forest fires and logging activities. Recently timber export to China has rapidly grown and is expected to accelerate degradation of forest resources in the Russian Far East. Based on statistical data and field surveys, timber export from the Russian Far East from the late 1980s to the present has rapidly increased, especially to China. Reasons for this rapid increase of exportation to China are the massive increase of timber demand in China due to economic growth and the recent Chinese policy protecting natural forests in order to preserve watersheds (Yamane 2007, 2008).

Russian forest policy is under major reform and is considered to have a significant influence on forest management and the timber industry. The policy system has been decentralized and the local government has become a key player in forest policy and management at the local level. Capacity building to policy formulation at the local level is an important emerging issue. Major reorganization of the national forest management system has been carried out and a large number of staff have been dismissed. There are strong concerns about the weakening of forest management ability at the field level (Kakizawa 2008).

The revision of the Forest Code in 2007 intends to support large scale companies which have the financial ability to invest in processing facilities. Concentration in the timber industry is expected to increase and to lead to an increase in log export taxes. More time is needed to assess how concentration in the timber industry will affect forest management.

Wetland conservation in Russia is not confined to the protection of Ramsar sites. Large wetland areas are conserved as part of the protected natural areas that were established domestically in Russia. Outside protected natural areas, wetland management is regulated by a number of laws (the Federal Law on the Conservation of the Natural Environment, the Federal Law on Environmental Impact Assessment, the Federal Law on Wildlife, and the Water and Forest Codes). However, there is still no efficient legal system which provides for an integrated solution to the various problems arising in the field of wetland use and conservation.

China Chinese forest management has long been under the strong control of the central government, while other environment regulations are within the jurisdictions of the Ministry of Environmental Protection or other inferior agencies. In 2001, a significant national initiative for the conservation of forests began and included the Natural Forest Protection Project, which provides for the prohibition of logging and the limitation of access in a large area of natural forest, and the Restoration of Cultivated Land to Forest Project, which purports to promote afforestation for the water retention function of forests. These projects are only partly successful, however, as they have resulted in the increase of the import of timber from the Russian Far East.

The wetland conservation, management and development in China involves more than ten governmental departments, including the Ministry of Environmental Protection, the Ministry of Agriculture, Ministry of Water Resources, the Ministry of Land and Resources and the State Forestry Administration. At present, no special national laws for wetland conservation are available in China. About 15 laws and regulations are applicable to wetland conservation including the Forest Law (1998 revision), the Land Administration Law (2004 revision), the Water Law (2002 revision) and the Regulation on Nature Reserves (1994).

At the provincial level, in particular in Heilongjiang Province, the conservation and management of wetlands have developed and been successful in recent years. To date, 28 Natural Reserves have been established in the Sanjiang Plain in Heilongjiang Province and comprise a total area of 5,958 km^2, in which three Natural Reserves have already been listed for Ramsar sites as wetlands of international importance under the Ramsar Convention.

The reclamation of the Sanjiang Plain reflects the historical background in which there has been active demand for food production and little understanding of the full significance of wetlands for environmental protection. China joined the Ramsar Convention in 1992, which is 21 years after the initial signing of the Convention in 1971. Most of the reclamation activities in the Sanjiang Plain took place before the 1990s. By the early 1990s, the perception of wetland functions was gradually growing in China. A large numbers of wetland reserves were established and wetland protection gradually became a mainstream environmental management concern.

In 1998, the Heilongjiang Province government issued the Decision about Strengthening the Conservation of Wetlands which stopped all reclamation and mining in the natural wetlands and tightened authority over development project in wetlands. Also since 1998, the project on conversion of croplands to wetlands in the plain had been implemented.

The Regulation about Conservation of Wetlands in Heilongjiang Province was issued for enforcement in 2003. Forestry departments at the county level and above are primarily responsible for the management of wetland resources. Under the regulation, a construction project which needs the use or reclamation of wetlands must be approved by the forestry department at the provincial level. The project should follow an environmental evaluation and a related approval process. Any actions damaging wetlands are illegal and any wrongdoers are punished.

At the national level, China has recently adopted some strict wetland management programs. For example, the National Wetland Conservation Action Plan (2000) is regulated under the leadership of the State Forestry Administration and is implemented by many relevant departments. Another is the National Implementation Program for Wetland Protection Engineering (2004), according to which 50% of the country's natural wetlands and 70% of its important wetlands were to be protected by 2010.

In some cases, the rules on wetland management in laws and administrative regulations overlap and conflict. For example, the Land Administrative Law (2004 revision) classifies most wetlands as a type of un-utilized land, and encourages organizations and individuals to exploit un-utilized land for use such as farming. As a result, under this law, most wetlands can be considered to be categories of land for which exploitation is encouraged (Wang et al. 2008a, b). On the contrary, many wetland protection programs require the conservation and the sustainable use of wetlands. In these circumstances, at least in legal terms, it is unclear which rule has priority over another.

For this reason, it is necessary to establish special laws that will standardize permitted and required behaviors for wetland protection and utilization, and that will coordinate existing laws so that they are compatible and people can use wetlands properly and consistently.

12.4 Future Conservation Strategy

Sections 12.2 and 12.3 demonstrate that the conservation of the GFBF ecosystem has been disrupted as a result of two human-made boundaries. In order to overcome these problems, the AOP proposed a new framework and strategy for the conservation of this ecosystem.

12.4.1 Amur-Okhotsk Consortium

While there is a tendency in this region to move in the direction of bilateral environmental cooperation, there has been no multilateral cooperative framework under which relevant countries can discuss the conservation of the GFBF ecosystem. Though an urgent threat to the ecosystem has not yet emerged, it has already become clear that the supply of dissolved iron from the Amur River basin regulates primary production in the Sea of Okhotsk and the Oyashio region and that land surface disturbances such as land-use change from wetlands to farmlands, forest fire and excessive loggings may impact the system in the future. In these circumstances, more data and information should be shared among the four countries concerning the extent to which such land surface disturbances can ultimately adversely affect primary production in the Sea of Okhotsk and the Oyashio region directly or indirectly. Without such information, regional cooperation for the fair distribution of the costs and profits is difficult.

Based on the precautionary principle, we suggest first promoting joint-monitoring, data exchange and mutual communication at the academic level among the four countries for the purpose of identifying the exact ecological conditions in the GFBF system more clearly. For this purpose, a group of academics and experts, including the AOP, established the "*Amur-Okhotsk Consortium*" in November 2009 as a multinational academic network to discuss the conservation and the sustainable use of the GFBF ecosystem.

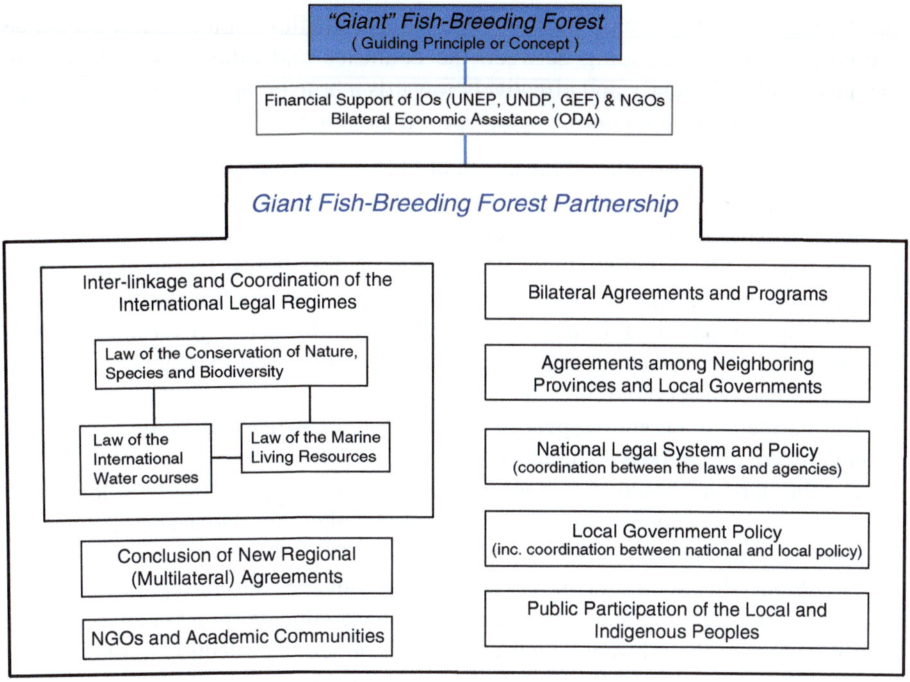

Fig.12.5 The structure of the GFBF partnership

The Consortium members comprise research institutes, universities, governmental organizations, international organizations and private companies. The Consortium will promote information sharing and exchange, and foster discussions on what must be done for the conservation of the Sea of Okhotsk and the Amur River basin. It is expected that such shared information and understanding will encourage each country to establish an intergovernmental regional framework for more robust cooperation in the future.

12.4.2 Institutional Design

Section 12.3 shows that while some environmental factors in the GFBF ecosystem have already been regulated by international and national laws and policies, these management regimes have been established and implemented independently; therefore, they are not adequate for the conservation of the whole GFBF system. It is concluded that it is important to coordinate and strengthen existing laws and policies in an integrated manner to manage this system consistently and effectively.

In order to achieve this coordination, the AOP proposed a multilateral institutional framework for the conservation of the GFBF system (Fig. 12.5). The aim of this framework is to develop common knowledge, facilitate information exchange, promote confidence-building between the countries and other stakeholders, and enhance work coherently and effectively towards a holistic approach to the conservation of the GFBF ecosystem, paying special attention to:

1. The "Giant Fish-Breeding Forest" should be recognized as a guiding principle, promoting the coordination, integration and reinforcement of existing international and domestic laws and policies.
2. Directed by this principle, the "Giant Fish-Breeding Forest Partnership" among the four countries can be built as a comprehensive flexible framework.
3. This Partnership should be established as a multi-layered governance system composed of multilateral, regional, bilateral, national and local level communications and conservation measures.
4. The Partnership should respect the bilateral agreements and programs already concluded among relevant countries.
5. The Partnership should encourage public participation, particularly of local people, indigenous peoples, non-governmental organizations and academic communities such as the "*Amur-Okhotsk Consortium.*"
6. The Partnership should facilitate international linkages and further development of existing international agreements that have already prescribed the basic obligations of the relevant countries for environmental protection in these areas. In addition, taking account of the ecological linkage between land and ocean in the GFBF system, rules, programs and institutions both in domestic and international dimensions should be coordinated coherently.
7. In this Partnership, relevant countries should conclude a new multilateral environmental agreement that requires these countries to facilitate mutual communication, negotiation and cooperation for the monitoring and assessment of the GFBF ecosystem, and prompts them to take measures for its conservation.
8. To promote consistency, other relevant measures which have been agreed or are being negotiated by some or all relevant countries in other forums should be take into account; in particular, the agreements between China and Russia concerning the joint-monitoring of the Amur River should be considered.
9. In order to provide a robust framework for policy formulation, financial and technical assistance from international organizations (ex. UNEP, UNDP, GEF), other countries and NGOs should be pursued for the effective and sustainable management of the GFBF system.

12.5 Concluding Remarks

The demarcation of national boundaries has usually been done on historical grounds, determined by political, economic and cultural factors, not environmental and ecological ones. State sovereignty has often been allocated to entities whose political borders disregarded the boundaries of the natural resources in question. At their worst, national borders were drawn with the colonizers' ignorance of geographic conditions in faraway continents (Benvenisti 2002). However, the effective management of ecosystems usually suffers when political boundaries do not take into account ecological and hydrological characteristics or correspond with natural units and ecosystems (Weiss 2007).

The fragmentation of environmental regimes, institutions and rules may be unavoidable to a certain extent, because the scope and purpose of each regime are always delimited without full scientific understanding of the ecosystems. In some cases, as scientific research has progressed, it has become clear that such circumstances may cause duplication or conflict between environmental regimes, finally resulting in the ineffective management of the ecosystems (Chambers 2008).

This paper has clarified that both the national boundaries among the four countries and the regime boundaries between various environmental laws and policies have disrupted the effective conservation of the GFBF system as a whole. Therefore, what is needed is to "blur" the strength of boundaries and mitigate the negative effects resulting from these two human-made boundaries. For this purpose, it is important to coordinate existing legal systems and policies in an integrated manner, and to establish an international cooperative framework that promotes a common understanding among the countries in this ecosystem.

Acknowledgements This work was done within the framework of the international project entitled as "Human activities in Northeastern Asia and their impact on the biological productivity in the North Pacific Ocean" funded by the Research Institute for Humanity and Nature (RIHN).

References

Benvenisti E (2002) Sharing transboundary resources: international law and optimal resource use. Cambridge University Press, Cambridge

Boyle A, Birnie P (2003) International law and the environment, 2nd edn. Oxford University Press, Oxford

Boyle A, Freestone D (eds) (1999) International law and sustainable development: past achievement and future challenges. Oxford University Press, Oxford

Brownlie I (1979) Legal status of natural resources in international law. In: Recueil des Cours: Collected Course of the Hague Academy of International Law, tome.162. Martinus Nijhoff, Leiden, pp 245–291

Chambers WB (2008) Interlinkages and the effectiveness of multilateral environmental agreements. United Nations University Press, Tokyo

FAO Investment Centre/EBRD Cooperation Programme (2008) Russian Federation: Review of the fishery sector: study supported under the Japan Europe cooperation, Report Series No. 12, FAO and EBRD

Faure M, Ying S (eds) (2008) China and international environmental liability: legal remedies for transnational pollution. Edward Elgar, Cheltenham

Hanamatsu Y et al (2010) The legal, political situations and a future conservation strategy of the giant fish-breeding forest. In: Shiraiwa T (ed) Report on Amur-Okhotsk Project, No. 6. Research Institute for Humanity and Nature, pp 87–104

Hassan D (2006) Protecting the marine environment from land-based sources of pollution: Towards effective international cooperation. Ashgate, Aldershot

Iwashita A (2005) An inquiry for new thinking on the border dispute: backgrounds of “historic success” for the Sin-Russian negotiations. In: Iwashita A (ed) Siberia and the Russian Far East in the 21st century: partners in the “community of Asia”. Slavic Research Center, Sapporo, pp 95–114

Kakizawa H (2008) Forest policy reform of Russian federation. In: Shiraiwa T (ed) Report on Amur-Okhotsk Project, No. 5, Research Institute for Humanity and Nature, pp 99–105

Kaul I, Grunberg I, Stern M (1999) Defining global public goods. In: Kaul I et al (eds) Global public goods: international cooperation in the 21st century. Oxford University Press, New York, pp 2–19

Kolossov V (2005) Border studies: changing perspectives and theoretical approaches. Geopolitics 10:606–632

McCaffrey S (2007) The law of international watercourses, 2nd edn. Oxford University Press, New York

Mensah TA (1999) The international legal regime for the protection and preservation of the marine environment from land-based sources of pollution. In: International law and sustainable development: past achievement and future challenges. Oxford University Press, Oxford, pp 297–324

Oberthür S, Gehring T (2006) Institutional interaction in global environmental governance: Synergy and conflict among international and EU policies. MIT Press, Cambridge

Onishi T (2007) Runoff properties of the Amur River and the construction of the hydrological model incorporating dissolved iron transport. In: Shiraiwa T (ed) Report on Amur-Okhotsk Project No. 4. Research Institute for Humanity and Nature, pp 201–206

Oude Elferink AG (2001) The sea of Okhotsk peanut hole: de facto extension of coastal state control. In: Stokke OS (ed) Governing high seas fisheries: the interplay of global and regional regimes. Oxford University Press, New York, pp 179–205

Querol M (2005) Rethinking international rivers and lakes as boundaries. In: De Chazournes LB, Salman SMA (eds) Water resources and international law. Martinus Nijhoff, Leiden, pp 97–132

Räsänen T, Laakkonen S (2008) Institutionalization of an international environmental policy regime: the Helsinki convention, Finland and the cold war. In: Joas M, Jahn D, Kern K (eds) Governing a common sea: Environmental policies in the Baltic sea region. Earthscan, London, pp 43–59

Sands P, Galizzi P (eds) (2004) Documents in international environmental law, 2nd edn. Cambridge University Press, Cambridge

Schrijver N (1997) Sovereignty over natural resources: Balancing rights and duties. Cambridge University Press, Cambridge

Shiraiwa T (2006) The Amur-Okhotsk project: trilateral cooperation to protect a shared environment. Gaiko Forum (Japanese Perspectives on Foreign Affairs) 6(3):36–42

Simonov E, Dahmer T (eds) (2008) Amur-Heilong River Basin Reader. Ecosystems, WWF, Hong Kong

UNEP (2006a) The Songhua river spill China, December 2005: Field Mission Report, UNEP. 26p

UNEP (2006b) Global international water assessment, GIWA Regional Assessment 30; Sea of Okhotsk, UNEP

UNEP (2006c) Global international water assessment, GIWA Regional Assessment 31, Oyashio Current, UNEP

Wang Y, Yao Y, Ju M (2008a) Wise use of wetlands: current state of protection and utilization of Chinese wetlands and recommendations for improvement. Environ Manage 41:793–808

Wang L et al (2008b) Conservation and alternative livelihoods: study on the Sanjiang plain wetlands. Social Science Academic Press, China

Weiss EB (2007) The evolution of international water law. Recueil des Cours: collected course of the Hague academy of international law, tome.331. Martinus Nijhoff, Leiden, pp 163–404

Wolfrum R, Matz N (2003) Conflicts in international environmental law. Springer, Berlin

Yamane M (2007) Recent development of forest resource use in Far East Russia, with a special focus on the relation to China. J Jpn Paper Assoc Pulp Paper 57(3):16–22, in Japanese

Yamane M (2008) Recent developments of the Sino-Russo timber trade in the Amur river basin. In: Shiraiwa T (ed) Report on Amur-Okhotsk Project, No. 5. Research Institute for Humanity and Nature, pp 107–116

Ying S (2008) International legal aspect of the Songhua River accident. In: China and international environmental liability: Legal remedies for transnational pollution. Edward Elgar, Cheltenham, pp 315–332

Zaisheng H, Jayakumar R, Liu Ke, Wang Hao, Chai Rui (2008) Review on transboundary aquifers in People's Republic of China with case study of Heilongjiang-Amur river basin. Environ Geol 54:1411–1422

Chapter 13
History of '*Uotsukirin*' (Fish-Breeding Forests) in Japan

Hiroshi Wakana

Abstract The oldest documented reference to '*uotsukirin*' (fish-breeding forests) in Japan is from the tenth century, and describes one in Tokushima Prefecture. *Uotsukirin* existed throughout Japan since the seventeenth century. Their main purpose was to promote sardine fishing, because sardines were an important industrial resource used to make commercial products including foods, lamp oil, and fish fertilizer at the time. In 1623, in the southern part of present-day Oita Prefecture, the first lord of the fief proclaimed, "I have heard that sardines do not come near the coast if the trees are not dense in the mountains along the seashore. To promote sardine fishing, therefore, I firmly forbid the felling of trees in small islands and the slash-and-burning in the mountains near the creek." He had become aware of the importance of the sardine in 1604. Huge amounts of sardines, herrings, and whales were transported into the inland as fish fertilizer in Japan in the seventeenth–nineteenth centuries, suggesting that the Japanese agriculture developed nationwide and on a large scale during the pre-modern Edo period (1603–1867).

In 1897, the Japanese government enacted the Forest Law. Japan's first protected forest as defined by a modern legal system was introduced in this Forest Law. In the twentieth century, research on the relationship between inland forests and the sea from the viewpoint of the natural sciences first appeared. Dr. Endo Kichizaburo, a professor at Sapporo Agricultural College who researched seaweed, described in 1903 that the cause of '*Isoyake*' (rocky-shore denudation) was the devastation of inland watershed forests. Dr. Inukai Tetsuo, professor of zoology at Hokkaido University, wrote in 1951 that forests all over Japan used to be *uotsukirin*. Nowadays, fishermen actively plant trees all around Japan.

Keywords Fish fertilizer • Forest law • Material cycle • Protected forest • *Uotsukirin* (Fish-Breeding Forest)

H. Wakana (✉)
Muroran Institute of Technology, 27-1 Mizumoto-cho,
Muroran, Hokkaido 050-8585, Japan
e-mail: wakana@mmm.muroran-it.ac.jp

M. Taniguchi and T. Shiraiwa (eds.), *The Dilemma of Boundaries: Toward a New Concept of Catchment*, Global Environmental Studies,
DOI 10.1007/978-4-431-54035-9_13,

13.1 Introduction

This article examines (1) the outlook of protected forests in Japan, (2) the relationship between industrial development and *uotsukirin* (fish-breeding forests) during the pre-modern Edo period (1603–1867), (3) scientific studies of *uotsukirin* in the twentieth century, and (4) recent tree planting movements initiated by local fishermen.

13.2 *Uotsukirin* and an Outline of Protected Forests in Japan

Under Japan's current Forest Law Article 25, there are 17 classified protected forests. Table 13.1 shows these classifications and the area of each as of March 2008.

The total area of the protected forests is 31.4% of the total area of Japan, with various protected forests comprising preservation networks.

The area of *uotsukirin* is 58,000 ha, or 0.5% of the total area of protected forests. Though this is not a large percentage, other protected forests, such as sandbreak,

Table 13.1 Total area of protected forests in Japan. Data are as of 31 March 2008

No.	Protected forest classification	Area (1000 ha)	Ratio (%)
1	[1] Water conservation forest	8,966	71.1
2	[2] Erosion control forest	2,506	19.9
3	[3] Landside prevention forest	57	0.5
Gross area of Nos. 1–3		11,529	91.5
4	[4] Sandbreak forest	16	0.1
5	[5] Windbreak protected forest	57	0.5
	[6] Flood damage prevention forest	1	0.0
	[7] Protected forest for tide break	13	0.1
	[8] Drought prevention forest	121	1.0
	[9] Snowbreak forest	0	0.0
	[10] Fog prevention forest	62	0.5
6	[11] Avalanche protected forest	19	0.2
	[12] Rockslide prevention forest	2	0.0
7	[13] Fire prevention forest	0	0.0
8	[14] Fish-breeding forest (*uotsukirin*)	58	0.5
9	[15] Navigation target forest	1	0.0
10	[16] Recreational forest	698	5.5
11	[17] Protected forest for scenic beauty	28	0.2
Gross area of Nos. 4–11		1076	8.5
[A] Gross area		12,606	100.0
[B] Net area		11,876	100.0
Ratio of [B] to total forest area of Japan		47.3%	
Ratio of [B] to total area of Japan		31.4%	

Note: Table translated from Japanese by the author based on data from the following website of the Japan Forestry Agency, Ministry of Agriculture, Forestry and Fisheries: http://www.rinya.maff.go.jp/seisaku/sesakusyoukai/tisan/syubetumenseki.htm (accessed September 21 2009)

Fig. 13.1 Explanatory signboard in Oita Prefecture. (Property of the Ministry of Environment and Oita Prefecture)

windbreak, snowbreak and tidebreak forests and a part of water conservation forests, have also recently been recognized as *uotsukirin.*

Figure 13.1 shows a signboard explaining the functions of the *uotsukirin* in the city of Saiki, Oita Prefecture. A translation of the content is as follows.

"*Uotsuki-hoannrin* (Fish-breeding Protected Forest): Many forests of the Tsurumi Peninsula are designated by the Prefecture as fish-breeding protected forests. These forests block direct sunlight and prevent sudden water temperature changes.

Moreover, when it rains the nutrients of the forests flows into the water, causing plankton to increase. Plankton is consumed by fish. Furthermore, the trees prevent the sea from becoming muddy even during heavy rains. Fish gather around such places, breeding and spawning."

Among the Japanese, the content of the signboard comprises common knowledge about *uotsukirin.*

Figure 13.2 shows an idealized pattern of reserved forests drawn by Endo Yasutaro in 1934. It depicts separate reserved forests being linked together in networks; with many reserved forests along the coast, river, and deep in the mountains. Endo drew 64 kinds of reserved forests in total.

Fig. 13.2 Idealized pattern of reserved forests. Endo (1934) drew 64 kinds of reserved forests (*foldout page*)

13.3 Spread of *Uotsukirin* in the Pre-modern Edo Period (1603–1867)

The oldest documented *uotsukirin* in Japan is recorded as being in Tokushima Prefecture in the tenth century. The location is near the present-day Tokushima Airport. Unfortunately, from the tenth to the sixteenth centuries, the locations of other *uotsukirin* remain unknown.

From the seventeenth century onward, however, *uotsukirin* spread throughout Japan. According to the available literature, *uotsukirin* were located in many places. Figure 13.3 is a historical map of *uotsukirin* in Japan, while Table 13.2 shows a

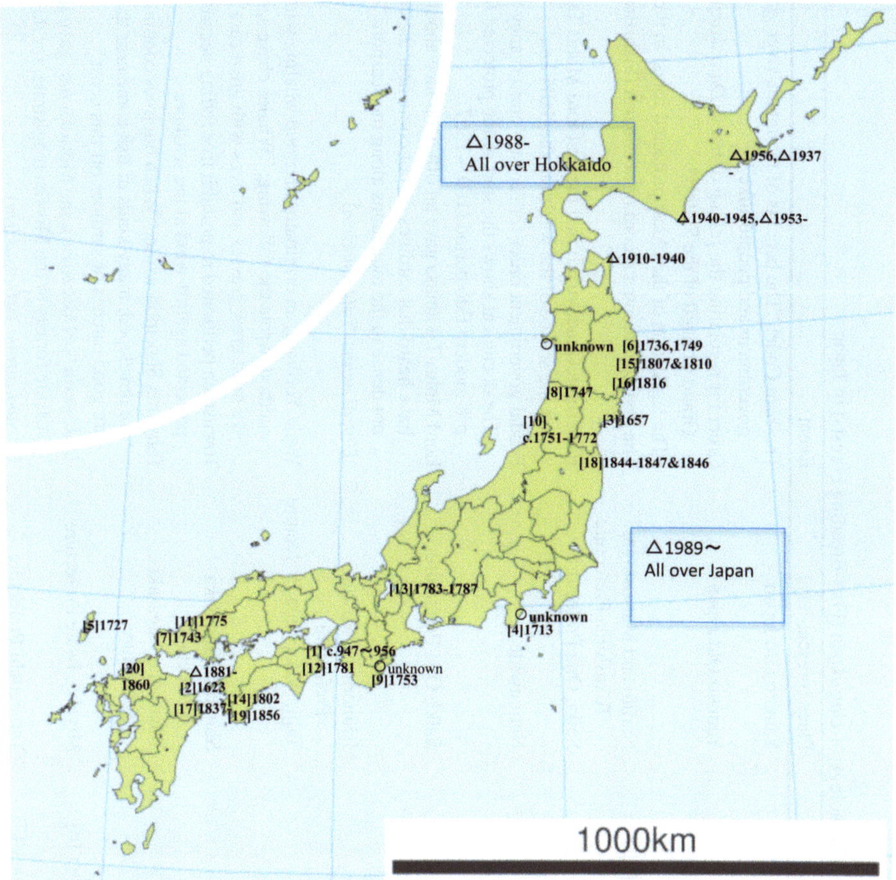

Fig. 13.3 Historical Map of *Uotsukirin* (Fish-Breeding Forests) in Japan *[1]–[20] and marks of *circle* unknown. *Uotsukirin* known to exist from 947 to 1860. ** marks of *triangle*: Projects to revive *uotsukirin* (since 1881)

Table 13.2 Chronology of *Uotsukirin* (Fish-Breeding Forests) in Japan

Year[a]	Place (present-day)	Event	Fish	Source[b]
718	Yamato:Old Japan	The *Yoro* Code: "The benefits of mountain, river ,grove , and marsh for government and people alike"		(1)
821	Yamato:Old Japan	Order for regulating the felling of trees (*Dajoukanpu* [Law of the Grand Council of the State])*		(1) (2)
		*The legislation of the oldest protected forest in the world		
c.947–956 [1]	Naruto and Matsushige, Tokushima Prefecture	The first documented record of *uotsukirin* (Fish-Breeding Forest)		(3) etc.
1604	Saiki, Oita Prefecture	Fisherman Gendayu advised the first lord Mohri Takamasa: 'Sardines are the source of the wealth in the Saiki *han*'	Sardine	(4)
1609	Nationwide	Feudal government order for forestry conservancy and flood control. The document shows the start of the protected forest system in the pre-modern Edo period (1603–1867)		(1)
1623 [2]	Saiki, Oita Prefecture	Lord Mohri Takamasa gave an official notice about *uotsukirin*': "I have heard that sardines do not come near the coast if the trees are not dense in the mountains along the seashore."	Sardine	(5)
1657 [3]	Ishinomaki, Miyagi Prefecture	'*Uotsukiyama* '(=*Uotsukirin*)		(3)
c.1675	Taiji, Wakayama Prefecture	Wada Yoriharu, an originator of group whaling with a traditional method, wrote the following: "Whales chase sardines, …,therefore, we must grew grasses and trees with great care"	Whale, Sardine	(6)
1713 [4]	Shimoda, Shizuoka Prefecture	Townsmen petitioned to prohibit tree felling because of fishing and for protected against wind at the seashore		(3)
1727 [5]	Tsushima, Nagasaki Prefecture	Tohyama Shoemon wrote, "When big trees continue to the seaside and are developed, many kinds of fish come near the shore; thus there are good catches of sardines off this coast."	Sardine	(3)
1736, October [6]	Miyako, Iwate Prefecture	'*Uokagerin*' (=*Uotsukirin*): the Morioka *han* gave the order to develop seaside forests to the Miyako magistrate, for the purpose of fishing	Salmon	(3) (7)
1743, February [7]	Yamaguchi Prefecture	'*Uotsukiajiroba*'(=*Uotsukirin*)		(3)

1747 [8]	Yamagata Prefecture	Planting post Hikozaemon and his son in the village of Shikano said, "If trees are planted in the seaside sand hills and the hills turn green, then fish will increase along the shore."	Sardine	(3)
1749	Miyako, Iwate Prefecture	Written record of oral statement: '*Sake-Tanekawa*' (protection of salmon breeding) System at the Tsugaruishi River estuary	Salmon	(8)
1753 [9]	Taiji and Nachi-Katsuura, Wakayama Prefecture. Kumano, Mie Prefecture	In the Wakayama *han*, because of poor fishing, disafforestation of the forest along the seashore was prohibited. There were *uotsukirin* in Taiji-Katsuura from previous times (date unknown). Tradition stated that villagers never broke a tree branch, and there was a custom of expelling those who cut a tree branch from the village. In Kumano City (Mie Prefecture), there were '*Ajiro-yama*,' '*Uotoriba-yama*', and '*Uotsuki-yama*' (=*Uotsukirin*), etc.	Whale	(3)
c.1751–1772 [10]	Murakami, Niigata Prefecture	Aoto Buheiji completed the '*Sake Tanekawa* ' (protection of salmon breeding) System. Since then, the felling of *Tabunoki* (Japanese persea) at the Miomote River estuary was strictly prohibited	Salmon	(9)
1775 [11]	Mukatsuku area in Nagato, Yamaguchi Prefecture	In Mukatsuku-kami village, two mountains are designated as '*Kujira-gyoba-uotsukiyama*' (=*Uotsukirin*)	Whale	(3)
1781 [12]	Komatsushima and Anan, Tokushima Prefecture	"At the seashore of Katsuura-gun Komatsushima-cho Kane-iso-shinden in the Tokushima *han*, a pine forest planted 30 years previously was designated as forest to protect the shoreline and called '*Kaisho-kuromi-gakari*'(=*Uotsukirin*). It was said that the forest was good for fishing. Moreover, at the Katsuura-gun, Naga-gun, and Kaihu-gun seashores, and in the case of '*Rodokoyama*' (=*Uotsukirin*), it was common to prohibit the felling of trees."		(3)
1783–1787 [13]	Higashiomi, Shiga Prefecture	"In the Hikone *han*, clansman Kusakabe Naiki began afforestation, and as a result of his efforts he developed a lush green *uotsukirin*. Fishing then became abundant, and the prosperity of the fishermen was promoted		(3)

(continued)

Table 13.2 (continued)

Year[a]	Place (present-day)	Event	Fish	Source[b]
1802 [14]	Uwajima , Ehime Prefecture	The Uwajima *han* gave the order: "Although sardines come near the seaside at *uotsukirin*, it is a poor catch all the more to do felling of the woods. And large quantities of seaweed have cropped to put the fertilizer on the terraced farms. It takes the breeding place of the fish by force, and it obstructs the propagation of the fish. ……You must not cut the tree of the mountain freely."	Sardine	(10)
1807, December. 1810, October [15]	Kamaishi, Iwate Prefecture	"The planting of *sugi* (cryptomeria) helps inshore fishery."		(3)
1816 [16]	Ofunato, Iwate Prefecture	"If *matsu* (Japanese pine) grows thickly, not only our village but also other villages will benefit. The *matsu* forest will become a navigational target forest and wind break forest and fish-breeding forest."		(3)
1837 [17]	Saiki, Oita Prefecture	Because reclaimed hills became barren, the amount of fishing decreased	Sardine	(11)
1844–1847 [18]	Soma, Fukushima Prefecture	Konno Bunzaemon of the Nakamura *han* said, "Many fish gather in areas where groves are near the seashore. But in sea near large sandy beaches which are not in the shade of trees and which are desolate, there are few fish."		(3)
1846, February [18]	Soma, Fukushima Prefecture	The Nakamura *han* gave the order to establish *uotsukirin:* "It is effective to plant trees along the seashore to (1) prevent farms and houses from sea breezes, (2) be useful for the army, and (3) provide good fishing."		(3)
1856, April [19]	Uwajima , Ehime Prefecture	In the Uwajima *han* village of Yusu, a grove of pines was left as '*Ajirokuromi*' (=*Uotsukirin*) because of fishing, and the felling of trees was prohibited	Sardine	(3) etc.
1860 [20]	Fukuoka Prefecture	In the Fukuoka *han*, an unknown person wrote, "Long ago, trees were planted in the mountain near creeks and islands. If the shadows of those mountains with trees are reflected into the sea, then fish approach the land. The planting of trees therefore makes an area teem with fish and shellfish."		(3)

1871, July	Nationwide	Abolition of the *han* (feudal domain) system and establishment of prefectures		
1881	Himeshima, Oita Prefecture	Himeshima Island became a barren island due to disafforestation. Chujo Ishitaro (the island leader) planted trees, causing the vegetation to revive, and from around 1907 fishing increased	Sardine	(11)
1897, April	Nationwide	Forest Law Article 8 Section 6 Enacted, establishing *uotsukirin* (fish-breeding forests)		
1903	Ito, Shizuoka Prefecture. Shima, Mie Prefecture. Tokushima Prefecture	Dr. Endo Kichizaburo (studies on seaweed, professor of Sapporo Agricultural College) wrote that the cause of *isoyake* (rocky-shore denudation) was the devastation of the inland watershed forests		(12)
1910–1940	Higashidoori, Aomori Prefecture	At the headland of Shiriya in Shimokita Peninsula, the youth association of Shiriya Village engaged in a project to afforest tree-fences to prevent sand movement; it was completed in 1940		(13)
1920	Shimokita Peninsula, Aomori Prefecture	As part of the Industrial Plan of Shimokita County, the author referred to complex functions of forest and the *isoyake* theory of Endo Kichizaburo		(14)
1937	Akkeshi, Hokkaido	Inukai and Nishio wrote 'Devastation of the inland forest caused the poor harvesting of oysters in Akkeshi Lake and Bay'.	Oyster	(15)
1940–1945	Erimo, Hokkaido	At the headland of Erimo, the Urakawa District Forest Office began afforestation, and stopped in 1945		
1951	Nationwide	Dr. Inukai Tetsuo (Zoology, Hokkaido University professor) wrote "All the forests in our country used to be *uotsukirin*."		(16)
1953, April	Erimo, Hokkaido	At the headland of Erimo, the Erimo Forestry Office was established. It began a project of Erimo afforestation		
1956	Akkeshi, Hokkaido	Began a project of 'Kon-sen Pilot-Forest', comprising of the afforestation of the riverhead area of Lake Akkeshi and Akkeshi Bay		(2)
1975	Hokkaido	Yaginuma, the leader of Hokkaido Federation of Fisheries Cooperative Associations, heard Inukai's issue: "Deep inland forests are more important for fish breending. All the forests in our country are *uotsukirin*"		(17)

(continued)

Table 13.2 (continued)

Year[a]	Place (present-day)	Event	Fish	Source[b]
1977	Erimo, Hokkaido	Erimo National Forest was designated as a recreational forest and *uotsukirin*. In 1961, it was designated as a sand break forest		
1988	Hokkaido	Conference of women's societies of the Hokkaido Federation of Fisheries Cooperative Association began movements of tree planting for the purpose of fish breeding		(18)
1989, September	Kesennuma, Miyagi Prefecture and Ichinoseki, Iwate Prefecture	At Murone Mountain in Iwate Prefecture, oyster farmers of Kesennuma Bay in *Miyagi* Prefecture started planting hardwood (a movement referred to as 'The Forest is the Sweetheart of the Sea'). Murone Mountain is the riverhead area of Kesennuma Bay	Oyster	(19)
2007	Northwest Asia	The concept of 'Giant' *Uotsukirin* is proposed		(20) (21)

In addition to the records shown, in the pre-modern Edo period (1603–1867), various coastal *han* (feudal fiefs) established *uotsukirin* throughout the country

[a][1]–[20] and show the position in historical map (Fig. 13.2)

[b]*Sources*: (1) Totman (1989) (Japanese tr. 1998), (2) Tomiyama (2001), (3) Endo (1934), (4) Village History of Yonohzu (1990), (5) Historical Materials of Saiki *Han*, Vol. 1(1995), (6) Tomiyama (2009), (7) Historical Materials of Miyako City, Vol. 6 (1990), (8) Sato (1986), (9) Yamaguchi (1957), (10) Miyamoto (2006), (11) Fishery Bureau, Ministry of Agriculture and Commerce (1911), (12) Endo (1903), (13) Village History of Higashidoori (Chapter History Part 2) (1999), (14) The Industrial Plan of Shimokita County (1920), (15) Inukai and Nishio (1937), (16) Inukai (1951), (17) Yaginuma (1993), (18) Yaginuma (1999), (19) Hatakeyama (1994), (20) Research Institute Humanity and Nature (2008), (21) Shiraiwa (2011)

chronology of events related to *uotsukirin*. Moreover, during the premodern Edo period (1603–1867), in addition to the following places, various coastal *han* (feudal fief) established *uotsukirin* in many more places. The following descriptions ([1]–[20]) were based on Endo (1934) along with local historical materials ([2] [6] [10] [14] and [17]) (see Fig. 13.3 and Table 13.2).

[1] Naruto and Matsushige, Tokushima Prefecture (947–956)
[2] Saiki, Oita Prefecture (1623): for sardines
[3] Ishinomaki, Miyagi Prefecture (1657)
[4] Shimoda, Shizuoaka Prefecture (1713)
[5] Tsushima, Nagasaki Prefecture (1727): for sardines
[6] Miyako, Iwate Prefecture (1736): for salmon
[7] Yamaguchi Prefecture (1743)
[8] Yamagata Prefecture (1747)
[9] Taiji and Nachi-Katsuura, Wakayama Prefecture, Kumano, Mie Prefecture (1753): for whales
[10] Murakami, Niigata Prefecture (1751–1772): for salmon
[11] Nagato, Yamaguchi Prefecture (1775): for whales
[12] Komatsushima and Anan, Tokushima Prefecture (1781)
[13] Higashiomi, Shiga Prefecture (1783–1787)
[14] Uwajima, Ehime Prefecture (1802): for sardines
[15] Kamaishi, Iwate Prefecture (1807)
[16] Ofunato, Iwate Prefecture (1816)
[17] Saiki, Oita Prefecture (1837): for sardines
[18] Soma, Fukushima Prefecture (1846)
[19] Uwajima, Ehime Prefecture (1856): for sardines
[20] Fukuoka Prefecture (1860)

The local names of *Uotsukirin* were various, as in the examples given below : *Uo-tsuki-yama* (1657, 魚著山,Miyagi Prefecture), *Uo-kage-rin* (1736, 魚蔭林, Iwate Prefecture), *Uotsuki-ajiro-ba* (1743, 魚附網代場, Yamaguchi Prefecture), *Ajiro-yama* (網代山) or *Uo-toriba-yama* (魚取場山) or *Uo-tsuki-yama* (魚附山) (1753, Wakayama Prefecture), *Kujira-gyoba-uotsuki-yama* (1775,鯨漁場魚附山, Yamaguchi Prefecture), *Kaisho-kuromi-gakari* (海上黒み懸り) or *Rodoko-yama* (艪床山) (1781, Tokushima Prefecture), etc.

These Japanese words mean the following: "*Uo*" (魚) = fish, "*tsuki*" (附•著•付•つき) = coming near, "*Yama*" (山) = mountain, "*kage*" (蔭) = shadow or (green) reflected light , "*rin*" (林) = forest, "*ajiro*" (網代) = wickerwork fish trap, "*gyoba*" (漁場) = fishing place, "*Uo-toriba*" (魚取場) = fishing place, "*Kujira*" (鯨) = whale, "*Kaisho*" (海上) = at sea, "*kuromi-gakari*" (黒み懸り) = with a tinge of black, "*Rodoko*" (艪床) = place of hanging scull in a Japanese ship.

In 1675, at Taiji, Wakayama Prefecture, Wada Yoriharu, originator of a group whaling with a traditional method, wrote the following: "Whales chase sardines, …,therefore, we must grow grasses and trees with great care."

These local names are based on the heuristic knowledge of various fishermen. I pay particular attention to words connected to light (e.g., "*kage*" (蔭) = shadow or (green) reflected light, "*kuromi-gakari*" (黒み懸り) = with a tinge of black). Ichthyologists have confirmed that fish are inclined to gather around the shining places with a specific spectrum of green reflected light, which can be seen as a tinge of black.

Uotsukirin spread throughout Japan from the seventeenth century, and it was against such a backdrop that sardine fishing was promoted. Sardines comprised important industrial resources for the production of foods, lamp oil, and fish fertilizer at the time.

In 1623, in present-day Saiki, the first feudal fief lord, Mohri Takamasa issued the following proclamation. "I have heard that sardines do not come near the coast if the trees are not dense in the mountains along the seashore. To promote sardine fishing, therefore, I firmly forbid the felling of trees in small islands and the slash-and-burning in the mountains near the creek."

He had become aware of the importance of the sardine in 1604 through the advice of a fisherman at Kamae Village named Gendayu. Specifically, Gendayu advised Takamasa, "Sardines are the source of the wealth in the Saiki *han* (feudal fief)."

Furthermore, in the eighteenth century, to promote salmon fishing and whaling, many lords encouraged the maintenance and expansion of *uotsukirin*.

As long as large numbers of salmon successfully ascended every year, the *han* could derive great benefit from salmon fishing. In 1736, the Morioka *han* gave an order to the Miyako magistrate to try to develop a seaside forest for the purpose of fishing (Iwate Prefecture).

From c.1751–1772, Aoto Buheiji (clansman of the Murakami *han*) completed the '*Sake-Tanekawa*' System, a system for protecting the spawning of salmon. From that time, the felling of the forest *Tabunoki* (Japanese persea) trees at the estuary of the Miomote River was strictly prohibited (Niigata Prefecture, Fig. 13.4).

The *han* derived great benefit from whaling as well. People made effective use of whales for foods, lamp oil, medicine, materials and fish fertilizer. The small chopped bones of whales, which were the oil cakes, were sent as fish fertilizer. In addition, whale oil was used for extermination of pests which affected rice cultivation. Whaling had drastically contributed to the development of Japanese rice cultivation since the eighteenth century (Ishida 1992).

Since the latter half of the eighteenth century, people had also used herring as a main source of fish fertilizer. Herring fish fertilizer was more effective and expensive than that of sardines.

Thus, from the seventeenth to nineteenth centuries, huge amounts of sardines, whales, and herrings were transported inland to be used as fish-based fertilizer in Japan, suggesting that the Japanese agriculture developed nationwide and on a large scale during the pre-modern Edo period (1603–1867).

Salmon carry nutrients from the sea to inland areas, and the bodies of salmon help the growth of trees. That is, the utilization of fish fertilizer represents the movement of sea nutrients to inland areas. Through such uses of fish fertilizer, the idea of *uotsukirin* spread throughout Japan during the seventeenth and nineteenth centuries.

Fig. 13.4 *Uotsukirin* signpost at the Miomote River estuary (Murakami, Niigata Prefecture). This *uotsukirin* began in the 1750s. It became officially designated as a protected forest in 1911 under the Forest Law, enacted in 1897

13.4 Projects to Revive *Uotsukirin* in the Twentieth Century

In 1871, the *han* system was abolished in Japan and was replaced with the prefectures system. In this transition, the major protected forest system (e.g., water conservation forests) remained basically unchanged, while the minor protected forests, including *uotsukirin,* entered a state of great flux.

The Japanese government enacted the Forest Law in 1897. This law included a revised protected forest system.

By 1881, Himeshima Island (Oita Prefecture) had become largely a bald mountain due to deforestation. To remedy this situation, Chujo Ishitaro (1847–1900), the island's leader, planted trees to increase the amount of vegetation on the island, and by 1907 fishing yield had increased (Fishery Bureau 1911).

From 1910 to 1940, at the headland of Shiriya in Shimokita Peninsula, the youth association of Shiriya Village engaged in a project to afforest tree-fences to prevent sand movement, with the afforestation being tentatively completed in 1940 (Higashidoori Village, Aomori Prefecture).

Similarly, from 1940 to 1945, at the headland of Erimo (Hokkaido), the Urakawa District Forest Office began afforestation efforts, but these actions were stopped in 1945 due to the aggravation of World War II. In 1953, the District established the Erimo Forestry Office to resume the Erimo afforestation project. In 1977 it designated Erimo National Forest as a recreational forest and *uotsukirin.* (It was also designated as a sand break forest in 1961).

13.5 Scientific Studies of *Uotsukirin* in the Twentieth Century

In the twentieth century, research on the relationship between inland forests and the sea from the viewpoint of the natural sciences first appeared. Dr. Endo Kichizaburo (1874–1921), a professor at Sapporo Agricultural College who research on seaweed, described in 1903 that the cause of '*Isoyake*'(rocky-shore denudation) was due to the devastation of inland watershed forests. The Sapporo Agricultural College (1876–1907) was a predecessor of the present Hokkaido University (1947–).

His theory was as follows. When heavy rains fall on a watershed's devastated forest land, a huge amount of fresh water spills into the bay, causing changes in the salinity distilled in the seawater. Such drastic changes kill seaweed.

Today, this theory is considered untenable. Nonetheless, his ideas represent the first natural science oriented research of the relationship between inland forests and the sea. His theory was introduced in the Shimokita County (Aomori Prefecture) industrial plan in 1920, which encouraged afforestation.

From 1931 to 1936, Dr. Inukai Tetsuo (1897–1989), professor of zoology at Hokkaido University, investigated the causes of oyster decreases in Lake Akkeshi (Hokkaido).

In 1937, Inukai and his colleague Nishio Shinroku reported their findings. They concluded that oysters breed in Lake Akkeshi from July to August. This time of year is the rainy season in Japan, and thus a huge volume of fresh water flows into Lake Akkeshi from the upper area of the Bekanbeushi River. In earlier times this was a primeval forest, but it had become barren by the time of the investigation. Under such conditions, the river carried large quantities of earth and sand into the lake, causing its water to become muddied whenever it rained. As a result, everything in the lake became covered in mud. Near a river's estuary, the temperature of lake water can change rapidly. In the case of oysters, such conditions mean that only healthy adult oysters can survive. Most shellfish fry die, resulting in a lower population of oysters. In this way, their study revealed that the devastation of forests in upstream regions had a great influence on the decline of the oyster population. Today, however, their hypothesis on the relationship between the growth of young oysters and the fluctuating temperature of lake water is also considered untenable. The currently prevailing view about the causes of oyster decreases is the overkill of the oyster and the devastation of inland forest areas.

Moreover, in 1951, Inukai wrote that forests throughout Japan could potentially be *uotsukirin* (1951). On his advice, the Japan Forestry Agency began the 'Kon-sen Pilot Forest' project in 1956, focusing on the afforestation of the riverhead area of Lake Akkeshi. This project is considered the largest afforestation project of the Showa period (1926–1989).

In 1975, Dr. Inukai lectured to fishermen of Hokkaido. Yaginuma Takehiko (1940–), the leader of the Hokkaido Federation of Fisheries Cooperative Associations, attended Inukai's lecture, too. Yaginuma heard Inukai's issue: "Deep inland forests are more important for fish breeding. All the forests in our country used to be *uotsukirin*". With this as a motivation, Yaginuma got the idea of initiating tree planting movements all over Hokkaido. With Yaginuma's suggestion, the Conference of women's societies of the Hokkaido Federation of Fisheries Cooperative Association started tree planting movements for the purpose of fish breeding from 1988.

Additionally, since 1989 fishermen throughout Japan have been actively planting trees.

13.6 Further Studies

Of course, fish captures fluctuate with various and variable factors. The existence of only *uotuskirin* cannot guarantee fish catches at all times. In fact, in the Japanese Islands, the captures of Japanese sardines (*Sardinops melanostictu*) have repeated cyclical fluctuations. The regime shift theory refers to the relation between sardine stocks and climate changes for several decades (e.g. Kawasaki et al. 2007; Kawasaki 2009). To mitigate these fluctuations, the Japanese fishermen had continued the longtime practice of *uotuskirin* to ensure continuous fishing. Meanwhile, scientists have conducted research into the effects of *uotuskirin* since the twentieth century. The Japanese fishermen's tree planting actions, in part, have resulted from the research and writings of Endo (1903) and Inukai Inukai (1951, 1937).

Until the nineteenth century, *uotsukirin* were restricted to waterfront areas, such as seashores and along rivers and lakes. In the latter half of the twentieth century, however, fishermen have successfully developed inland woodland areas into *uotsukirin*.

Even today, the actions of Japan's fishermen influence both national and local government policies.

As the twenty-first century unfolds, scientists have proposed the concept of "Giant" *uotsukirin* in the seas of Northwest Asia (see Chap. 8: "Giant" Fish-Breeding Forest: A New Environmental System Linking Continental Watershed with Open Water).

References

Endo K (1903) *Kaisan-shokubutsugaku* (Marine Botany), *Hakubunkan* (in Japanese)

Endo Y (1934) *Nihon sanrinshi* (History of mountains and forests in Japan), vol 1, *Nihon-sanrinshi-kankoukai* (in Japanese)

Fishery Bureau, Ministry of Agriculture and Commerce (1911) *Nihon no Uotsukirin* (*Uotsukirin* in Japan), Chapter of Oita Prefecture, Shinzansha (in Japanese)

Hatakeyama S (1994) *Mori ha umi no koibito* (The Forest is sweetheart of the sea), *Hokuto-shuppan* (in Japanese)

Higashidoori Village (1999) *Higashidoori sonshi rekishi-hen 2* (Village history of Higashidoori, Chapter History Part 2) (in Japanese)
Inukai T, Nishio S (1937) A limnological study of Lake Akkeshi with special reference to propagation of the oyster. J Facul Agr Hokkaido Imper Univ 40:1–33
Inukai T (1951) *Shinrin to suisangyou* (Forest and fisheries), *Juhyo*, pp 1–3 (in Japanese)
Ishida Y (1992) *Rettou no hogeibunka* (Culture of whaling in the Japanese Islands), Umi to rettoubunka (The Sea and Culture in the Japanese Islands), vol 10, *Shogakukan* (in Japanese)
Kawasaki T, Hanawa K, Taniguchi A, Nihira A (eds) (2007) *Rejgime shift: kikouhendou to seibutusu sigen kanri* (Regime shift climate changes and biological resources management), *Seizando-shoten* Publishing (in Japanese)
Kawasaki T (2009) *Iwashi to kikouhendou* (Sardine and climate changes), *Iwanami-shinsho* (in Japanese)
Miyako City Board of Education (ed) (1990) *Miyako-shishi shiryoushu kinsei 6* (Historical materials of Miyako city, early modern history, vol 6) (in Japanese)
Miyamoto H (2006) *Danpata to Iwashi karano Kotozute* (1) (Messages from terraced farms and sardines vol 1), *Souhuusha-shuppan* (in Japanese)
Research Institute Humanity and Nature (2008) Brochure 2008, pp 12–13 (in Japanese)
Saiki City Board of Education (ed) (1995) *Saiki-han shiroyo Onko-chishin dai 1 kan* (Historical Materials Saiki han, vol 1) (in Japanese)
Sato S (1986) *Sake* (Salmon), *Iwanami-Shinsho* (in Japanese)
Shimokita County (1920) *Shimokitagun no sangyo keikaku-an* (The Industrial Plan of Shimokita County), 68pp (in Japanese)
Shiraiwa T (2011) *Uotsukirin no Chikyukankyougaku* (global environmental studies on fish-breeding forest), *Showado* (in Japanese)
Tomiyama K (2001) *Kankyo-mondai toha nanika* (What is the environmental problem?), *PHP-shinsho* (in Japanese)
Tomiyama K (2009) (*Shizen to Ningen*) *Umi ha ikiteiru* ([Nature and Human] The seas are alive), Kodansha (in Japanese)
Totman C (1989) The green archipelago, forestry in preindustrial Japan, University of California Press (Japanese translation 1998)
Wakana H (2001) The development of idea of '*uotsukirin*' (fish-breeding forests) in the modern age in Japan. J Water Environ Issues 14:1–9 (in Japanese)
Wakana H (2004) The beginning of the idea of '*uotsukirin*' (fish-breeding forests) and its material cycle in Japan. J Water Environ Issues 17:53–62 (in Japanese)
Wakana H (2008) Afforestation movement in Shimokita peninsula and Sapporo agricultural college. J Water Environ Issues 21:1–14 (in Japanese)
Yaginuma T (1993) *Ki wo uete sakana wo huyasu* (Tree planting for fish breeding), *Ienohikarikyokai* (in Japanese)
Yaginuma T (1999) *Mori ha subete uotsukirin* (All forests are fish-breeding forest), *Hokuto-shuppan* (in Japanese)
Yamaguchi K (1957) *Nihon gyogyo-shi* (History of Fisheries in Japan), *Tokyo-daigaku-shuppankai* (The University of Tokyo Press) (in Japanese)
Yonohzu Village (1990) *Yonohzu-sonshi* (Village history of Yonohzu) (in Japanese)

Chapter 14
Integrating Groundwater Boundary Matters into Catchment Management

William Todd Jarvis

Abstract With 97% of the world's freshwater resources stored underground, the connection between groundwater resources to the metrics of space, scale and time common to the geographic study of natural resources has not been extensively investigated by political geographers. Recognized as a common pool resource, the management and governance of groundwater resources is challenging and increasingly conflictive not only due to its hidden nature, but also because of the difficulty in placing boundaries around the groundwater resources and user domains.

Given that groundwater is the world's most extracted raw material with withdrawal rates estimated to range between 800 and 1,000 km^3 per year through millions of water wells, the groundwater domain boundaries are three-dimensional and change with time. A previously unrecognized typology for groundwater resources and user domains determined that (1) traditional approaches to defining groundwater domains focus on predevelopment conditions, referred to herein as a "commons" boundary, (2) groundwater development creates human-caused or a "hydrocommons" boundary where hydrology and hydraulics are meshed, and (3) the social and cultural values of groundwater users define a "commons heritage" boundary acknowledging that groundwater resources are part of the "common heritage of humankind". This typology helps define a fundamental unit of analysis to aggregate demographic, social, and economic data. Emerging paradigms of groundwater governance suggest "unitizing" some groundwater development situations as one means to mitigate the inefficiency of a possession or use-based system of groundwater along with the inefficiencies associated with joint access to groundwater. Yet drawing these domain boundaries is supremely political and morphs with changing

W.T. Jarvis (✉)
Department of Geosciences, Institute for Water and Watersheds,
Oregon State University, 210 Strand Hall, Corvallis, OR 97331, USA
e-mail: todd.jarvis@oregonstate.edu

M. Taniguchi and T. Shiraiwa (eds.), *The Dilemma of Boundaries: Toward a New Concept of Catchment*, Global Environmental Studies,
DOI 10.1007/978-4-431-54035-9_14, © Springer Japan 2012

social and cultural values. Incompatibilities often arise over the use and equitable, or inequitable, distribution of groundwater, "values" attached to groundwater, conceptual models, uncertainty, as well as on missing information, inaccurate data, and how the "science" will be used by knowledge entrepreneurs.

Keywords Boundaries • Governance • Groundwater • Groundwater domains • Problemshed • Unitization

14.1 Introduction

Groundwater is a resource that is found everywhere. Groundwater use is increasing because it is a "commons" resource, available to anyone with the financial resources to drill, equip, and power a well (Moench 2004). Pumping of groundwater is among the most intensive human-induced changes in the hydrologic cycle. With dramatic changes in drilling technology, pumping technology and the availability of electrical and diesel power over the past 60 years, the number of wells has increased exponentially in many parts of the world (Moench 2004; Shah 2009). According to Zekster and Everett (2004) and Shah (2009), groundwater is the world's most extracted raw material, with withdrawal rates approaching 800–1,000 km^3 per year through millions of water wells.

The global economy is becoming increasingly dependent on groundwater. Just as the "radius of influence" associated with pumping wells continues to grow with time, the economic impact of groundwater use quickly moves from the point of use to the world economy just like the model of territories and scale in the world system as described by Kolossov and O'Loughlin (1998). For example, more than 90% of the groundwater pumped from the Ogallala Aquifer located in central North America is used to irrigate crops for food and fiber worth more than $20 billion US dollars per year to global markets (Little 2009). Bottled water from springs is shipped thousands of kilometers, fueling a global industry worth nearly $50–100 billion US dollars per year (Glennon 2002).

Environmental flows and ecosystem services are more dependent on groundwater than previously thought. Dai et al. (2009) estimated that continental discharge including runoff from all land areas except Antarctica and Greenland approaches 37,288 km^3 per year. Gautier (2008) estimates that 36% of river runoff comes from groundwater, or approximately 13,425 km^3 per year. As more groundwater is captured by wells, the result is either induced recharge primarily from surface water crossing the recharge area or decreases in discharge to surface water systems (Bredeheoft 1997). Likewise, virtually all groundwaters constitute ecosystems ranging from microbes to larger species depending on the permeability architecture of the groundwater system (Gibert et al. 1994).

Information power or the "infosphere" of Lonsdale (1999) is perhaps the most important dimension of strategy in the geopolitics of groundwater. While groundwater monitoring networks have been in place in many intensively exploited regions of the world, few have been expanded since their inception (Moench 2007). Groundwater data collection and information dissemination has created a large industry of "knowledge entrepreneurs" as described by Conca (2006).

Despite this growing infosphere, with 97% of the world's freshwater resources stored underground, the connection between groundwater resources to the metrics of space, scale and time common to the geographic study of natural resources has not been extensively investigated by political geographers. The goal of this chapter is to provide an overview of groundwater boundaries within catchments, particularly the social and political boundaries, which do not necessarily line up with the technical or physical hydrologic boundaries.

14.2 The Boundaries of Groundwater

Despite the substantial body of geographic literature surrounding the historical, cultural and political development of boundaries (e.g., Anderson 1999), it is ironic that few political geographers have addressed the problem of how boundaries are placed around common pool resources such as groundwater. Clearly, the new world order of groundwater will focus on the delineation of resource and user domain boundaries regardless if (1) the technological options to manage groundwater quantity and quality problems employ water transfers, managed recharge, or conjunctive use, or (2) the resource governance solutions include (a) collective or community action, (b) developing instrumental approaches such treaties, agreements, rights, rules and (c) prices or other incentives such as preserving the structural and ecological integrity of groundwater systems as summarized by Giordano (2009).

14.2.1 The Boundary Conundrum

Resource domains define the fixed spatial dimensions of resources (Buck 1998). Fish stocks, for example, are natural resources found in the ocean resource domain. Spatial dimensions are used to define property rights which may be held by individuals, groups of individuals, communities, corporations, or nation-states. Rights to natural resource properties are not a single right, but are rather composed of a "bundle of rights" such as rights of access, exclusion, extraction, or sale of the captured resource; the right to transfer rights between individuals, communities, corporations or nation-states; and the right of inheritance (Buck 1998). Each "right" has an implied boundary. The spatial extent of a resource affects both the ability of users to develop information and to assess their relative ability to capture the benefits of organization (Schlager 2007). Yet with the assumptions associated with the bundle of rights and implied boundaries comes the fact that the assumptions, knowledge and understandings that underlie the definition of the rights and associated boundaries are uncertain and often contested (Adams et al. 2003). For example, the question of identity and its relation to the domains of natural resources is often overlooked (Dietz et al. 2002). Choices about water resources are value choices that involve distinct local communities of interest (Blomquist and Schlager 2005). Defining boundaries around water resource domains is "a supremely political act" because these boundaries represent different interpretations of key issues such as water quality, water quantity, nature, economics and history (Blomquist and Schlager 2005). The resulting boundaries may range from the international scale, to the national, regional, local, or even the individual scale. These come from the fact that water resources are coupled with the larger reality of a region, including its environmental, social, legal, and economic characteristics.

14.2.2 Why Boundaries Matter

Boundaries are "inner-oriented" or created by the will of a central government, or two or more states in an international setting, with the boundary indicating the limits of a political unit. All that falls within the confines of the boundary has a common bond. According to Casati et al. (1998), possession of a boundary is one mark of individuality in the ontology of geographic representation. Without a boundary, there can be no separation and control, and without control, it is doubtful where sovereignty in the full sense can be enjoyed (Bisson and Lehr 2004). The existence of a boundary is the first criterion for the individuality of an autonomous entity.

Consideration of a transdisciplinary and broad systems approach to exploring the geopolitics of groundwater yields a typology for groundwater boundaries. As depicted in Fig. 14.1, this work found that (1) traditional approaches to defining groundwater domains focus on predevelopment conditions, referred to herein as a bona-fide "commons" boundary, (2) groundwater development creates human-caused or *fiat* "hydrocommons" boundary where hydrology and hydraulics are meshed, and (3) the social and cultural values of groundwater users define a *fiat* "commons heritage" boundary acknowledging that groundwater resources are part of the "common heritage of humankind". The significance of this typology is that it focuses more on "problemsheds"—the boundaries of a particular problem defined by the issue-network—than on watersheds or catchments as recommended for water governance and management by Molden (2007).

Boundaries that correspond to physical differentiations or spatial continuities in territories constitute natural or bona-fide boundaries. Coastlines, rivers, watersheds or catchments, or rock outcrops are good examples of bona-fide boundaries for groundwater resource domains.

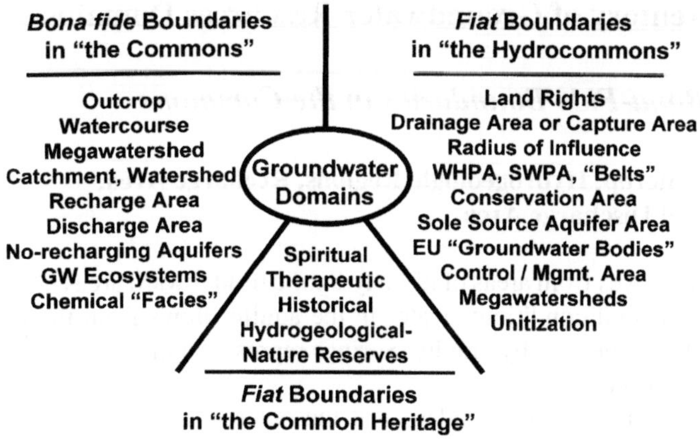

Fig. 14.1 Inventory of groundwater domains

Fiat boundaries are subjective boundaries demarcated by humans based on judgment and "ease" and represent groundwater user domains. Borders between countries are *fiat* boundaries. According to Anderson (1999), boundaries have no horizontal dimension, as the crucial dimension of boundaries lies in the vertical plane or subsurface beneath the boundary. Three dimensional *fiat* objects are created by subterranean volumes of land assigned rights to minerals, the ocean, or groundwater. The capture area of a wellfield is an example of three-dimensional *fiat* boundaries in groundwater user domains (Casati et al. 1998).

The spatial representation of groundwater resource domains in geographic terms varies across space, scale, time and depth. The spatial, temporal and boundary aspects of groundwater resource domains are important in defining water budgets or the sustainable yield of groundwater resources (Maimone 2004; Moench 2004). Boundaries can represent water lost or gained from over- or underlying aquifers, areas of direct recharge, areas of subsurface discharge to coastal areas or lakes and discharge to streams as base flow.

Groundwater resource domains are a nested series of spatial and temporal configurations (Gibert et al. 1994). The megascale domain represents the regional through continental groundwater flow systems that can extend over tens to hundreds of kilometers and depths ranging from 800 m to 3 km (Garven 1995; Bisson and Lehr 2004). At the macroscale, geomorphologic and hydrologic processes of catchments and watersheds determine aquifer properties, the permeability architecture, and water circulation characteristics. The mesoscale domain incorporates hydrodynamic controls, matter and energy fluxes, and human impacts. Human impacts by intensive exploitation of groundwater for drinking water, irrigation, and energy development are important at this scale. The microscale domain represents short-term events such as during the annual hydrologic cycle and at a spatial scale of the pore, fissure or channel (Gibert et al. 1994).

14.3 Inventory of Groundwater Resources Domains

14.3.1 Bona-Fide Boundaries in the Commons

14.3.1.1 Outcrop, Hydrogeologic Regions, Recharge Area, and Discharge Area

Groundwater moves from areas of high hydraulic head toward areas of low hydraulic head. Under natural conditions, water in the aquifer flows from the recharge area toward the discharge area by gravity. Springs represent a discharge area, or areas of low hydraulic head.

The significance of the boundaries of the outcrop and regional hydrogeologic maps from the perspective of groundwater resource domains focuses on (1) a first-order approximation of what areas recharge the groundwater resource domain, and (2) the storage characteristics of the groundwater system. Recharge is the process by which groundwater is replenished. A recharge area is where water from precipitation is transmitted into an aquifer. Areas which transmit the most recharge into a groundwater system are often referred to as "critical" recharge areas with unique boundaries for land use management. Conversely, discharge areas are locations at which groundwater leaves the aquifer and flows to the surface as springs or water bodies such as wetlands, streams, rivers, lakes or the sea.

At the global scale, the International Groundwater Assessment Center (IGRAC 2009) developed a map of global groundwater regions that differentiated 35 regions on the basis of tectonic setting, present-day geomorphology, and the spatial extent of rock formations with contrasting hydraulic properties as part of the consortia of institutions undertaking the World-wide Hydrogeological Mapping and Assessment Programme (WHYMAP) (2008). Building upon the International Groundwater Resources Assessment Center (IGRAC) mapping, the WHYMAP further refined the hydrogeologic regions into hydrogeologic units that form over 270 transboundary aquifers. The boundaries of the groundwater resources domains are based primarily on permeability architecture. For example, when viewed on a global scale, groundwater basins with sedimentary rocks compose 35%, complex hydrogeologic regions compose 18%, and shallow aquifers typically associated with Precambrian igneous and metamorphic rocks compose 47% of the aquifer types in the world (WHYMAP 2008). It is clear that there are multiple approaches to defining a groundwater resource domain and that scale is an important factor when defining such a domain.

14.3.1.2 Watercourses

The 1997 Convention on the Non-Navigational Uses of International Watercourses defines a watercourse as "a system of surface waters and groundwaters constituting by virtue of their physical relationship a unitary whole and flowing into a common terminus" (McCaffrey 2001:34). While a watercourse can be conceptualized as a watershed associated with a river, a watercourse can also chiefly consist of groundwater, where precipitation within the recharge zone may not be necessarily associated with a surface stream, with the terminus of the watercourse dependent on the local geology of the groundwater system. The terminus of the "groundwater course" may be the receiving aquifer, a related aquifer, or the sea (McCaffrey 2001:25).

14.3.1.3 Watershed, Catchment, and Drainage Basin

The watershed, catchment, or drainage basin of a river is the boundary with a long history serving as the boundary for water resources management, particularly within the world's 270 international river basins (Bakker 2009; Wolf and Giordano 2002). The International Law Association drafted the Helsinki Rules of 1966 and the Seoul Rules of 1986 which identified the international drainage basin as the unit to delineate the geographic extent of surface water and groundwater considered under the rules (Matsumoto in Delli Priscoli and Wolf 2009).

14.3.1.4 Megawatersheds

Bisson and Lehr (2004) define the conceptual model of a megawatershed as a natural complex system of water catchments and drainages linked to tectonism, consisting of three-dimensional surface and subsurface zones linking interbasin groundwater transmission in consolidated fractured bedrock and compartmentalization in faulted, deep sediments. As mappable groundwater resources, megawatersheds may not coincide with surface topography divides, and they may receive recharge from parts of several surface catchments.

Within the western United States, the carbonate rocks composing the Great Basin Aquifer serve as an excellent example of a deeper megawatershed due to the apparent lack of hydraulic connection between groundwater flows in the nearly 260 surface watersheds that overlie the Great Basin Aquifer System. The Great Basin Aquifer underlying the states of Utah, Nevada, Idaho, and Oregon are targeted for development by the Southern Nevada Water Authority to provide groundwater supplies approaching 0.21 km^3 per year to Las Vegas (Bredehoeft and Durbin 2009).

14.3.1.5 No-Recharging Aquifers or Nonrenewable Groundwater

Declining water levels suggest that groundwater resources may be "nonrenewable". The term "nonrenewable groundwater" is controversial because it refers to groundwater resources where present-day replenishment is limited but aquifer storage is large, where replenishment is very long (hundreds to thousands of years) relative to the time-frame of human use, or where the use of groundwater storage is at a rate much greater than the renewal rate, essentially "mining" the groundwater or "decoupling" it from the hydrologic cycle due to changes in the climatic conditions in the watershed

(Foster and Loucks 2006). Areas of nonrenewable groundwater can be found on every continent where rainfall is less than 200 mm per year (WHYMAP 2008). According to Zekster and Everett (2004), deep irregularly-recharged aquifers or "fossil" groundwater constitute relevant water management regions because the groundwater systems are not hydraulically linked to a watercourse.

14.3.1.6 Groundwater Ecosystems

The hydraulic interaction between groundwater and aquatic ecosystems has long been recognized by groundwater hydrologists. In 1972, the United States implemented Section 404 of the Federal Clean Water Act (CWA) which initiated the process of protecting wetlands from pollution and destruction. The Water Framework Directive (WFD-2000/60/EC) for the European Community also recognized the need to protect aquatic ecosystems. Both the CWA and the WFD emphasize terrestrial ecosystems. However, virtually all groundwaters constitute ecosystems ranging from microbes to larger species depending on the permeability architecture of the groundwater system (Gibert et al. 1994). There also exists the need to delineate the boundaries of groundwater ecosystems (Stanford and Gibert 1994).

14.3.1.7 Chemical Facies

A groundwater resource domain may also be an area where good quality groundwater is wholly or partially surrounded by poorer quality groundwater. Poorer quality groundwater can be divided into two groups: naturally occurring problems such as high salinity, arsenic, and fluoride, and areas of groundwater contaminated by agricultural, industrial, military, sewage, and municipal wastes (Moench 2004; Giordano 2009).

14.3.2 Fiat Boundaries in the Hydrocommons

Focusing primarily on rivers and watersheds, Weatherford (2003) defined the "hydrocommons" as the convergence of hydrology and hydraulics yielding an area defined by the linkages of common water sources. Whether a hydrocommons represents a grounded reality remains debatable, as Weatherford (2003) opines that the "Hydrocommons seems less a reality than a metaphor for the fragmentation of natural resource planning and management". Yet this fragmentation results in a broad spectrum of user domains, particularly when groundwater resources are concerned.

14.3.2.1 Land Ownership Rights

Land administration systems traditionally focus on rights to surface ownership under the rules of Roman Law, Napoleonic Civil Code, and English Common Law.

In some countries, land ownership includes not only the ground surface, but also all earth layers below, including all groundwater. The "rule of capture" presides over groundwater ownership in these settings, where the groundwater user is permitted to pump as much water as can be physically captured.

14.3.2.2 Radius of Influence and Capture Areas

Wells drilled into aquifers "capture" water stored in the aquifer by creating artificial discharge areas by lowering the hydraulic head in the vicinity of the well during pumping (Theis 1940). When a well is pumped, the water level is drawn down in the immediate vicinity of the well. This drawdown is referred to as the cone of depression in the vertical dimension; in the planimetric dimension the cone is represented as the "radius of influence". Groundwater flowing toward the well during pumping, which is derived from storage in the aquifer, is called the "capture area". The size of the capture area is not only a function of the hydraulic properties of the aquifer, but also the pumping rate and the duration of pumping.

The geographic significance of both the radius of influence and the capture area is that these can become management areas, requiring policies directing how many wells can be drilled into an aquifer to efficiently exploit the water stored in the aquifer, or protect from overdrafting (Livingstone et al. 1996). The intersections of radii of influence associated with pumping multiple wells can be compared to the intersection of "water rights domains".

14.3.2.3 Drainage Areas

Large quantities of groundwater are developed as a by-product of mine dewatering or oil and gas development that can lead to extensive areas of drainage. Likewise, large areas can be drained of groundwater by horizontal water "mines" such as qantas or karezes, found in the tens of thousands throughout the Middle East, which capture groundwater via gravity and drain towards portals.

14.3.2.4 Wellhead Protection Areas and Source Water Protection Areas

A wellhead or source water protection area as defined by nearly every state and province in the United States and Canada, as well as some countries in Europe, is the surface and subsurface area around a well, spring, or tunnel through which contaminants are reasonably likely to move toward and contaminate the drinking water source. Two domains are commonly used to describe the areas near a well: (1) a fixed radius near the well to protect the area in the immediate vicinity of the well, and (2) the zone of contribution, or "capture area" determined from sophisticated computer models, to hydrogeologic mapping using remote sensing supplemented by groundwater tracing with dyes or isotopes (Livingstone et al. 1996).

14.3.2.5 Sole Source Aquifer Areas

The Sole Source Aquifer (SSA) program was established under Section 1424(e) of the U.S. Safe Drinking Water Act of 1974. The SSA designation authorizes the U.S. Environmental Protection Agency (EPA) Administrator to assess that an aquifer is the "sole or principal source" of drinking water for an area. An aquifer must supply 50% or more of the drinking water for an area to qualify as "sole or principal" (McCabe et al. 1997).

14.3.2.6 Groundwater Bodies

The Water Framework Directive (WFD) produced by the European Commission to direct the achievement of sustainable management of waters in the European Union Member States. According to the Working Group on Water Bodies, the WFD covers all waters, including surface water, groundwater, transitional, and coastal waters up to one sea mile form the territorial baseline of a Member State. The geographical or administrative unit for water management is the river basin or river basin district. Groundwaters are associated with a river basin or river basin district.

A body of groundwater within the WFD refers to a distinct volume of groundwater within an aquifer or aquifers. The WFD does not provide explicit guidance on how the bodies of groundwater are delineated. The groundwater bodies can be identified as (1) separate within different strata overlying each other in the vertical plane, or (2) single bodies within the different strata. The final approach to defining the groundwater bodies is up to the individual Member States, but it must be assigned to a River Basin District.

14.3.2.7 Conservation Areas

Article VII of the Bellagio Model Agreement Concerning the Use of Transboundary Groundwaters proposed in 1989 indicates that Transboundary Groundwater Conservation Areas can be determined by the Commission of the Agreement (Hayton and Utton 1989). While the model agreement is silent with respect to the standards used to develop the boundaries of the conservation area, the emphasis is on (1) the sustained use of the groundwater resource by groundwater withdrawals exceeding recharge to endanger yield, water quality, or diminish the water quantity or quality of interrelated surface water; (2) the impairment of drinking water; or (3) the contamination of aquifer(s).

14.3.2.8 Legislative Boundaries

In many parts of the United States where large withdrawals of groundwater occur or where water quality has been impaired over large areas, a broad spectrum of tools for local management of groundwater have been developed. Some of these areas have been the result of court orders, others are legislative mandate, and others are created voluntarily. The legislative bodies have a plethora of names such as Control Areas, Aquifer Authorities, Management Areas, Natural Resource Districts, and Water User's Group. The boundaries of the districts are generally developed along political boundaries with little regard for the geologic or hydrologic boundaries of the groundwater systems.

14.3.2.9 Megawatersheds

Just as catchments are reshaped, breached and bounded by hydraulics resulting in hybrid surface water resource domains as defined in the "hydrocommons", the extra-basin area enclosing the collection and distribution of imported water to a groundwater resource domain is also part of the modern hydrocommons. Aquifer storage and recovery and aquifer replenishment programs that transmit surface water hundreds of kilometers from distant river basins to intensively exploited groundwater basins is becoming increasingly commonplace in the United States (Pyne 2005). Injection wells, subsurface dams or "sea cutoff walls" used to control saltwater intrusion in coastal aquifers effectively expand the boundaries of the groundwater user domain oceanward (Sakura et al. 2003). The net effect is an expansion of the hydrocommons or the creation of a megawatershed.

14.3.2.10 Unitization

Economists and legal scholars in property rights suggest "unitizing" some situations associated with groundwater development as one means to mitigate the inefficiency of a possession or use-based system of groundwater along with the inefficiencies associated with joint access to groundwater. Unitization is a common practice in the oil industry to preserve the structural integrity of a reservoir. In the case of groundwater, Libecap (2005) defined the concept in the following way: "... government-mandated unitization of groundwater ... is a solution to excessive access and drawdown ... [where] a single "unit operator" extracts from and develops the reservoir. All other parties share in the net returns as share holders." The boundary issue is critical to the success of unitization (Kumar 2007). Unitization may serve as one option to dealing with the emerging issue of overlapping jurisdictions or "values" related to groundwater.

14.3.3 *Fiat Boundaries in the Commons Heritage*

Besides the natural or physical boundaries of a groundwater resource domain or the user domain boundaries derived from the exploitation of groundwater, there are natural and human boundaries associated with groundwater. The intensive use of groundwater is causing increased awareness of potential impacts to groundwater dependent ecosystems, spiritual resources, therapeutic resources, cultural and historical resources, and geothermal resources. Many of these resource domains could be considered part of the common heritage of humankind or part of the global commons to which all nations have legal access as opposed to international commons which are resource domains shared by several nations (Buck 1998).

14.3.3.1 Nature Reserves

In order to protect deep confined aquifers, as well as spring waters and mineral waters used as therapeutic waters as part of a national or common heritage, de Marsily (1994) calls for the creation of "Hydrogeological Nature Reserves". Springs are a concentrated discharge of groundwater at the ground surface. The recharge area drained by a spring is referred to as a springshed. Springer and Stevens (2009) propose 12 "spheres" of spring discharge and associated microhabitats.

14.3.3.2 The Special Case of Thermal Springs and Springsheds

Hydrothermal features are nature reserves that are not frequently considered within the context of groundwater resource domains. Yet, many have been used for thousands of years for therapeutic purposes, are World Heritage Sites, national and state parks, and are spiritually important to some cultures. For example, the Sipapuni is a geothermal spring located near the Grand Canyon National Park and flows approximately 0.3 L per second. The Sipapuni is considered the place of emergence for Hopi ancestors from the Third World to the Fourth World (Dongoske et al. 1997).

Hydrothermal features are increasingly being explored as sources for renewable energy. For example, the world's largest mineral hot springs located at Thermopolis, Wyoming and Yellowstone National Park in the U.S. are at risk of "running out of steam" due to groundwater capture from nearby wells (Prevost 2006). Discoloration of the cotton white travertine terraces at the World Heritage Site at Pamukkale, Turkey, by diversion of the thermal waters for tourist hotel pools, coupled with the lack of sewage systems leading to algae growth on the terraces, has led to conservation efforts (Simsek et al. 2000). Boundaries for groundwater user domains associated with common heritage sites vary from no protection to controlled areas.

14.3.4 The Third Dimension

Up to this point, the discussion has focused on the planimetric boundaries associated with groundwater domains under both static and dynamic conditions. Lopez-Gunn and Jarvis (2009) synthesized the results of conceptual and numerical models of ideal groundwater systems, along with the mechanical limitations associated with pumping groundwater from great depths, to argue for a vertical dimension for distinguishing between shallow and deeper groundwater systems. Their work proposed a groundwater governance model that differentiates between shallow groundwater systems and deep groundwater systems. They found that shallow groundwater systems that are hydraulically connected to surface water resources may extend to depths approaching 305 m. This boundary was proposed as a starting point for differentiating between groundwater that may be governed under surface water regimes and deeper groundwater systems that may be governed as part of the global commons.

14.4 Discussion and Conclusions

Changes in population, the world's climate, effectiveness of water treatment and conservation technologies, and social values all affect the rate of groundwater pumping, recharge, and ecological response with time. The types of spatial entities associated with groundwater resource domains not only occupy space, but also sometimes share space with other spatial entities. Governance, management, and use of shared resources such as groundwater cannot be treated independently and require viewing the linkages between the user and resource domains as a problemshed rather than through the lens of the catchment (Molden 2007).

Boundaries, either political or defining a resource or user domain, are obviously related to the control or the distribution of groundwater; boundaries are used to exclude some users while at the same time providing the appropriators an opportunity to develop information and capture the benefits of organizing within the boundaries (Schlager 2004, 2007).

Social psychologist Mark van Vugt (2009) identified four conditions for the successful management of shared resources: information, identity, institutions, and incentives. Placing boundaries around user and resource domains (1) helps reduce the uncertainty within the groundwater "infosphere" and decreases reliance on knowledge entrepreneurs, (2) builds social identity and organization with groundwater resources, (3) localizes the institutional controls to promote fairness and trust regarding the use of groundwater, and (4) provides a roadmap to incentives to preserve the integrity of an aquifer or the associated ecosystem services. Yet global reviews of water-related conflicts by Blomquist and Ingram (2003) and Thomasson (2005) determined that the root causes of such conflicts include limited resources, control of distribution, and the quality of the resource, all of which have related boundaries. Clearly, boundaries can create competition between competing communities and institutions that do not promote the welfare of groundwater resources. Van Vugt (2009) suggests that it is important to think of ways to "blur" the boundaries by promoting the outlook that "we are all in this together" in order to reduce conflict and competition as described by Jarvis and Wolf (2010). The Sole Source Aquifer, unitization, and hydrogeologic nature reserves outlined herein provide opportunities to blur the boundaries when it comes to protecting groundwater quality, quantity, and spirituality by providing holistic approaches to integrating a watershed approach to groundwater protection, by acknowledging that all groundwater users are shareholders in a groundwater system, and by recognizing that multiple uses of water extend beyond simply the commodification of water.

References

Adams WM, Brockington D, Dyson J, Vira B (2003) Managing tragedies: understanding conflict over common pool resources. Science 302:1915–1916

Anderson EW (1999) Geopolitics: international boundaries as fighting places. In: Gray CS, Sloan G (eds) Geopolitics, geography, and strategy. Frank Cass, Portland

Bakker MHN (2009) Transboundary river floods: examining countries, international river basins and continents. Water Pol 11:269–288

Bisson RA, Lehr JH (2004) Modern groundwater exploration: discovering new water resources in consolidated rocks using innovative hydrogeologic concepts, exploration,aquifer testing, and management methods. Wiley Interscience, Hoboken

Blomquist WA, Ingram HM (2003) Boundaries seen and unseen: resolving transboundary groundwater problems. Water Int 28:162–169

Blomquist WA, Schlager E (2005) Political pitfalls of integrated watershed management. Soc Nat Resour 18:101–117

Bredeheoft JD (1997) Safe yield and the water budget myth. Ground Water 35:929

Bredehoeft J, Durbin T (2009) Ground water development—the time to full capture problem. Ground Water 47:506–514

Buck SJ (1998) The global commons: an introduction. Island Press, Washington

Casati R, Smith B, Varzi AC (1998) Ontological tools for geographic representation. In: Guarino N (ed) Formal ontology in information systems. Ios Press, Amsterdam

Conca K (2006) Governing water: contentious transnational politics and global institution building. MIT Press, Cambridge

Dai A, Qian T, Trenberth KE, Milliman JD (2009) Changes in continental freshwater discharge from 1948 to 2004. J Climate 22:2773–2792

De Marsily G (1994) Hydrogeological nature reserves? future groundwater resources at risk. IAHS publication no 222. Centre for Ecology and Hydrology, Oxfordshire

Delli Priscoli J, Wolf A (2009) Managing and transforming water conflicts. Cambridge Press, New York

Dietz T, Dolsak N, Ostrom E, Stern PC (2002) The drama of the commons. In: Ostrom E, Dietz T, Dolsak N, Stern PC, Stonich S, Weber EU (eds) The drama of the commons. National Research Council, National Academy Press, Washington

Dongoske KE, Yeatts M, Anyon R, Ferguson TJ (1997) Archaeological cultures and cultural affiliation: Hopi and Zuni Perspectives in the American southwest. Am Antiq 62:600–608

Foster S, Loucks DP (eds) (2006) Non-renewable groundwater resources: a guidebook on socially-sustainable management for water-policy makers IHP-VI. Series on groundwater no 10. United Nations Educational Scientific and Cultural Organization, Paris

Garven G (1995) Continental-scale groundwater flow and geologic processes. Annu Rev Earth Planet Sci 23:89–117

Gautier C (2008) Oil, water, and climate: an introduction. Cambridge University Press, New York

Gibert J, Stanford JA, Dole-Oliver MJ, Ward JV (1994) Basic attribute of groundwater ecosystems and prospects for research. In: Gibert J, Danielpol D, Stanford J (eds) Groundwater ecology. Academic, San Diego

Giordano M (2009) Global groundwater? issues and solutions. Anns Rev Environ Resourc 34:7.1–7.26

Glennon R (2002) Water follies: groundwater pumping and the fate of America's freshwaters. Island Press, Washington

Hayton RD, Utton AE (1989) Transboundary groundwaters: the Bellagio draft treaty. Nat Res Jour 29:677

International Groundwater Resources Assessment Centre (IGRAC) (2009) Transboundary aquifers of the world. http://www.igrac.net/publications/320. Accessed 23 August 2010

Jarvis T, Wolf A (2010) Managing water negotiations and conflicts in concept and in practice. In: Earle A, Jägerskog A, Öjendal J (eds) Transboundary water management: principles and practice. Earthscan, London

Kolossov V, O'Loughlin J (1998) New borders for new world orders: territorialities at the *fin-de-siecle*. GeoJournal 44:259–273

Kumar MB (2007) Geological aspects of unitization in the petroleum fields of Louisiana: a brief overview. The Professional Geologist Nov–Dec: 30–32. http://dnr.louisiana.gov/assets/docs/TPG_Kumar-Articles.pdf Accessed 09 February 2012

Libecap GD (2005) The problem of water, essay prepared for national bureau of economic research. http://www.aeaweb.org/annual_mtg_papers/2006/0108_1300_0702.pdf. Accessed 27 June 2010

Little JB (2009) The Ogallala aquifer: saving a vital U.S. water source. Sci Amer Earth 3.0 19:32–39

Livingstone S, Franz T, Guiger N (1996) Managing ground-water resources using wellhead protection programs. Geoscience Canada 22:121–128

Lonsdale DJ (1999) Information power: strategy, geopolitics, and the fifth dimension. In: Gray CS, Sloan G (eds) Geopolitics, geography, and strategy. Frank Cass, Portland

Lopez-Gunn E, Jarvis WT (2009) Groundwater governance and the law of the hidden sea. Water Pol 11:742–762

Maimone M (2004) Defining and managing sustainable yield. Ground Water 42:809–814

McCabe WJ, Job CA, Simons JJ, Graves JS, Terada CJ (1997) History of the sole source aquifer program: a community-based approach for protecting aquifers used for drinking water supply. Groundwater Monit Rev Summer 1997:86

McCaffrey SC (2001) The law of international watercourses, non-navigational uses. Oxford University Press, New York

Moench M (2004) Groundwater: the challenge of monitoring and management. In: Gleick P (ed) The world's water 2004–2005. Island Press, Washington

Moench M (2007) When the well runs dry but livelihood continues: adaptive responses to groundwater depletion and strategies for mitigating the associated impacts. In: Giordano M, Villhoth KG (eds) The agricultural groundwater revolution: opportunities and threats to development. CABI International, Oxfordshire

Molden D (ed) (2007) Water for food, water for life. Earthscan/International Water Management Institute, London/Colombo

Prevost R (2006) Is Thermopolis out of steam? Casper, WY.: the casper star-tribune, march. http://trib.com/news/state-and-regional/article_68042562-9beb-50e1-93d7-5d6567b17a46.html. Accessed 27 June 2010

Pyne RDG (2005) Aquifer storage and recovery: a guide to groundwater recharge through wells, 2nd edn. ASR Systems, Gainesville

Sakura Y, Tang C, Yoshioka R, Ishibashi H (2003) Intensive use of groundwater in some areas of China and Japan. In: Custodio E, Llamas R (eds) Intensive use of groundwater: challenges and opportunities. AA Balkema, Leiden

Schlager E (2004) Common-pool resource theory. In: Durant RF, Fiorino DJ, O'Leary R (eds) Environmental governance reconsidered: challenges, choices, and opportunities. MIT Press, Cambridge

Schlager E (2007) Community management of groundwater. In: Giordano M, Villhoth KG (eds) The agricultural groundwater revolution: opportunities and threats to development. CABI International, Oxfordshire

Shah T (2009) Taming the anarchy: groundwater governance in south Asia. Resources for the Future Press, Washington

Simsek S, Gunay G, Elhatip H, Ekmekci M (2000) Environmental protection of geothermal waters and travertines at Pamukkale, Turkey. Geothermics 29:557–572

Springer AE, Stevens LE (2009) Spheres of discharge of springs. Hydrogeol J 17:83–93

Stanford JA, Gibert J (1994) Conclusions and perspective. In: Gibert J, Danielpol D, Stanford J (eds) Groundwater ecology. Academic, San Diego

Theis CV (1940) The source of water derived from wells-essential factors controlling the response of an aquifer to development. Civ Eng 10:277–280

Thomasson F (2005) Local conflict and water: addressing conflicts in water projects. Swedish water house. http://www.swedishwaterhouse.se/swh/resources/20050425162906Water_and_Local_Conflict.pdf. Accessed 23 August 2010

van Vugt M (2009) Triumph of the commons: helping the world to share. New Sci 2722:40–43

Weatherford GD (2003) Out of the basin, into the hydrocommons. In: Utton Transboundary Resources Center (ed) Conference report and synthesis on interstate waters crossing boundaries for sustainable solutions, a multidisciplinary approach. University of New Mexico School of Law, Albuquerque

Wolf AT, Giordano MA (2002) Atlas of international freshwater agreements. Early warning and assessment report series, RS. 02-4. United Nations Environment Programme, Nairobi

World-wide Hydrogeological Mapping and Assessment Programme (WHYMAP) (2008) Groundwater resources of the world. http://www.whymap.org/cln_116/whymap/EN/Home/whymap__node.html?__nnn=true. Accessed 27 June 2010

Zekster IS, Everett LG (2004) Groundwater resources of the world and their use, IHP-VI, series on groundwater no 6. Springer, Dordrecht

Chapter 15
A Boundary Between Surface Water and Groundwater in Japanese Legal System: Its Consequences and Implications

Takahiro Endo

Abstract The concept of the hydrologic cycle has significance not only for the natural sciences but also for the social sciences. As a result of this cycle, an upstream water diversion may place external diseconomies on downstream water uses. A conflict of interest can result. A well-designed institution overseeing water resource management would have a function to prevent or resolve such externality problems. This paper focuses on the influence of the artificially created boundary on this function. The investigation is based on a case study of a groundwater problem in Saijo City, Ehime, Japan. While surface water is defined as "public water" and is subject to governmental regulations, groundwater is regarded as "private water" and goes with the land ownership under the Japanese legal system. This boundary between surface water and groundwater, which does not exist in the natural hydrological cycle, can hinder internalization of external diseconomies between surface water and groundwater users. It is often vaguely asserted that integrated management of surface water and groundwater is necessary for integrated water resources management. This paper shows that efficient use of water resources provides a basis for that assertion.

Keywords Externality • Integrated water resources management • Private water • Public water • Saijo City

T. Endo (✉)
Graduate School of Life and Environmental Sciences, University of Tsukuba,
1-1-1 Tennoudai, Tsukuba, Ibaraki 305-8572, Japan
e-mail: endo@envr.tsukuba.ac.jp

M. Taniguchi and T. Shiraiwa (eds.), *The Dilemma of Boundaries: Toward a New Concept of Catchment*, Global Environmental Studies,
DOI 10.1007/978-4-431-54035-9_15, © Springer Japan 2012

15.1 Introduction

The concept of the hydrologic cycle has significance for both natural and social sciences. This cycle connects people with each other through their use of water. A person's water use often impacts on another person's. This characteristic often causes conflicts over water resources, especially when water is scarce.

When an actor directly influences another without any compensation, such an influence is called an externality. If it is a positive one, it is categorized as an external economy; if a negative one, it is an external diseconomy (Scitovsky 1954).[1]

The reason why externality matters is that it can mean an inefficient use of resources. A conflict between upstream and downstream users of water is a typical example. An upstream user tends to take as much water as it wanted without thinking of any negative impact on the downstream user. That is because the upstream user is apt to underestimate the total cost of a diversion and does not calculate the cost that directly accrues to the downstream user. If the (marginal) benefit to the former is less than the social (marginal) cost, overall welfare worsens. This results in an inefficient use of a resource, so that if the former changes the way the water is used (i.e., diverts less), there is room for the overall welfare to be improved. Theoretically, the same thing can be applied not only to conflicts between riparian users, but also to conflicts over a shared aquifer. This framework also fits into the cases involving water quality issues.

A well-designed institution for water resource management would be expected to have a function to prevent significant externalities on water utilization. If there is a rule that forced a person causing an external diseconomy to compensate for the damage, he/she will take the full cost of the water use into consideration. Then the increased cost would lead to less diversion and would improve social welfare. This is what we call internalization of the externality (Mankiew 1998).

The focus of this paper is on the influence of human-made boundaries on this function. There are various water-related boundaries that hinder internalization. A country upstream in an international river does not always recognize the negative impacts that affect the countries downstream. National borders often hinder information sharing and cooperation between riparian countries (Berber 1959; Elhance 1999). Further boundaries can be found inside a country, for example, in regards to sectionalism within a government. Fragmented policy-making between waterworks and wastewater treatment may create a situation where a series of wastewater outlets are located above drinking water intake facilities. The biological effects of dam construction are often ignored when the administration of water quantity and quality are fragmented. As these examples show, human-made boundaries often become obstacles to effective water management.

---

[1] In this paper, externality is defined as technical externality; meaning that an actor's activity impacts another actor and does not pass through to the market. That is, it is an effect that the former has on the latter without permission or compensation. Thus it is distinguished from a pecuniary externality. For this definition, see Scitovsky (1954).

From among these examples, this paper deals with a boundary that has so far been paid little attention; i.e., a legal boundary between surface water and groundwater. In Japan, while surface water is generally regarded as "public water" and is subject to many public regulations imposed under national law (the River Law), groundwater is basically treated as "private water" and belongs to the overlying land so that a landowner can make free use of it.[2] What kind of role does this legal boundary have in promoting the efficient use of water resources? In this paper, the presence of diseconomies will be demonstrated through a case study on a groundwater problem occurred in Saijo City. Some implications regarding water management will be deduced from this case.[3]

15.2 Groundwater Problem in Saijo

In Japan, groundwater, being a common source of irrigation and drinking water, has historically been used for a variety of purposes. After 1950, however, many dams were built and people became primarily dependent on surface water: groundwater only accounts for about 12.6% of total present water use. Nevertheless, groundwater is still an important source of water in some regions, and remains a primary reserve water supply in case of drought. As a consequence, there has been a renewal of interest in groundwater management as climate change becomes a social issue (Ministry of Land, Infrastructure, Transport and Tourism 2008).

Saijo City, Ehime, Japan, is a small city in Shikoku-Island with a population of about 58,000 people in the year 2000 (Saijo City 2003; Fig. 15.1). The city (covering an area of approximately 231 km^2) is located on an alluvial fan created by the flow of the Kamo River that originates from the Ishizuchi mountain area. The city is comprised of a mountainous area (73%) and a plain (27%) that faces the

[2] There are multiple criteria used for defining "public water." (1) Whether it is running water, (2) Whether it is navigable, or (3) Whether it has something to do with public interest or not, and so on (Ichimura 1909). The River Law defines "River" as "public running water and water surface" and regards it as public water (Tanaka 1968). Not all rivers in Japan are subject to the River Law. The law applies only to rivers classified as First or Second class. The rest are managed by the corresponding application of the River Law (the River Law §3, §100, Study Group on the River Law, 2006).

[3] As mentioned detailed below, the groundwater problem in Saijo illustrates a conflict of interest between Saijo and its adjacent city, Matsuyama regarding the water in the Kamo River. The latter wants to divert water from Kamo River but the former is very anxious because the diversion will most probably decrease the river flow and have a negative impact on groundwater recharge for groundwater users in Saijo City. The proposal is still under negotiation. Moreover, the possible on groundwater in Saijo has yet to be quantitatively estimated. Thus the problem we discuss is just a potential one that caused by the different legal statuses of surface water and groundwater. But those legal distinctions are applied throughout Japan. Even if the problem is not yet actual in Saijo, similar problems can take place elsewhere. In this sense, the case of Saijo could be an example of future issues. This is a reason why Saijo was chosen as the case study model for this investigation.

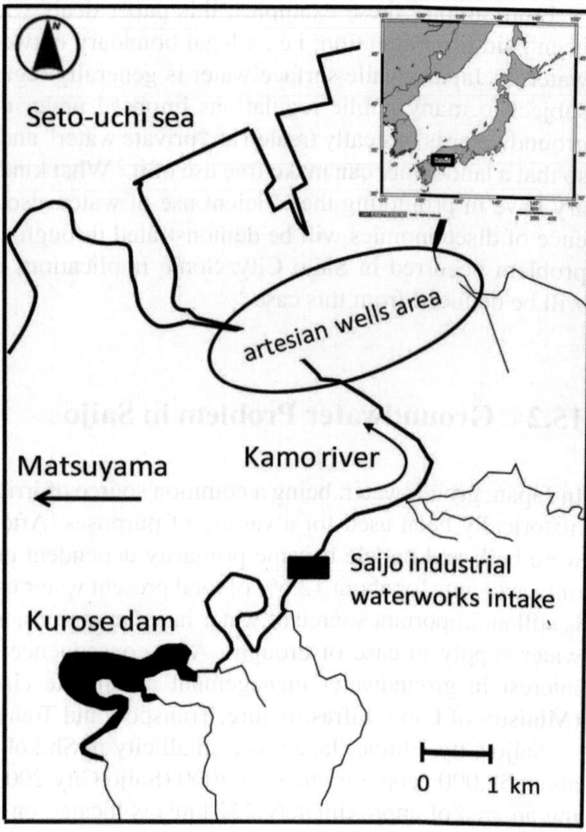

Fig. 15.1 Map of Saijo City

coast: annual precipitation was 2,294 and 1,334 mm, respectively, based on records from 1969 to 1998 (Saijo City 2000).[4]

Saijo has a long history of groundwater use. It is recorded that a team of well-digging craftsman operated as early as in 1842 (Miki 1993). Even today, groundwater is used for many purposes such as irrigation, industrial, and domestic uses. Figure 15.2 shows the population and tap water users in Saijo and its neighboring areas. The Saijo residential area totally depends on groundwater for its tap water supplies. This figure shows only 24% of residents get their water from waterworks. The rest get water directly from wells called Uchinuki. Groundwater is literally the basis of daily life in Saijo City (Saijo City 2001).

It is assumed that surface water (Kamo River) is one of the main sources of groundwater recharge (Suzuki et al. 1998). Saijo City government once estimated that 85.8% of the groundwater in the Saijo Plain came from seepage from the river

[4] By way of contrast, the average precipitation per year for Japan, over a 30 year measurement period (1976–2005), is about 1,700 mm. (Ministry of Land, Infrastructure, Transport and Tourism 2008).

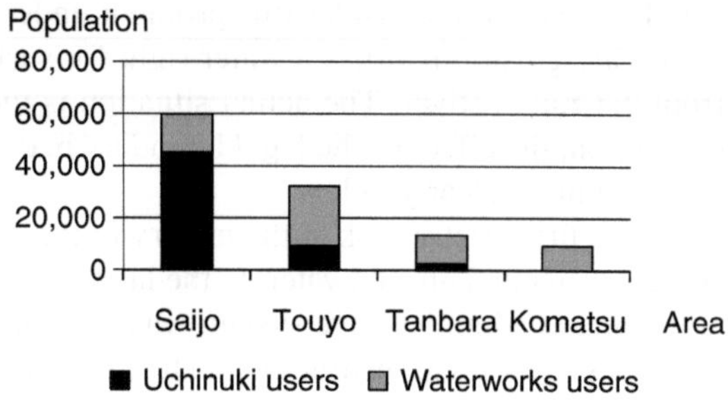

Fig. 15.2 Population and waterworks users in Saijo and neighboring areas. *Source*: Saijo City (2010)

(Saijo City 2000). Thus, it is thought that securing the flow of the Kamo River is crucial for residents of Saijo City.

A conflict over groundwater use is presently taking place because of a water diversion plan put forward by Matsuyama City government, located 50 km west of Saijo City. Matsuyama City government has established new water resource development as a primary policy objective after it experienced a severe drought in 1994. The municipality considered various means to secure a stable water supply and concluded that water diversion from the Saijo Industrial Waterworks (SIW) was the best.[5] Because the intake and storage facilities of SIW are installed in the Kamo River, this plan equates to a water diversion plan for the Kamo River. Saijo City government is against this plan because it would decrease the river flow and harm groundwater supplies in Saijo City, especially in drought years. Using the concept of externality, it is clear that Saijo City government is anxious that the diversion plan will create an external diseconomy which will affect the residents of Saijo city.

It should be noted that this diversion plan is still under consideration and not implemented as of November 2010. Moreover, the impact that could occur on groundwater has not yet been quantitatively estimated. Thus it is impossible to say that Saijo City government's anxiety will be realized; however, it cannot be denied that a potentially harmful effect exists.

15.3 The Legal Boundary Between Surface Water and Groundwater and Its Effect

From a theoretical viewpoint, the conflict between Matsuyama City and Saijo City is one between surface water and groundwater users. How can the Japanese legal system resolve this externality problem between these two kinds of water users?

[5] Originally, these waterworks were built and run by Ehime Prefecture to promote industrial development in the coastal areas of Saijo, and an adjacent city, Niihama. But, as a result of industrial changes caused by economic depression in early 1970's, the number of factories built were so limited that the demand for industrial water was far less than expected. The SIW suffered a chronic fiscal deficit (Official Website of Bureau of Public Utilities, Ehime Prefecture).

It will be useful to consider this problem under a hypothetical situation where the Saijo City government has a water right for surface water and takes water directly from the Kamo River. The actual situation is then discussed. With this intentional distinction, the effect of the legal boundary between surface water and groundwater is shown more clearly below.

Let us first briefly look at the history of the Japanese legal system as it relates to surface water and groundwater. In the latter half of the nineteenth century, a modern political and administrative system was established in Japan in line with European examples. New legal institutions related to water resources also took shape.

As for the development of rules on groundwater, it is said that two factors had dominant influences. First, the Japanese Supreme Court decision of 1896 (Kanazawa and Sanbongi 1979) established the precedent that the right of use of groundwater went with the ownership of the land and a landowner could make free use of groundwater lying below his/her land (*Japanese Supreme Court, Civil Case, Book II*, vol. 3 (in Japanese)). The second is the Japanese civil law, which went into effect just 1 year after the Supreme Court judgment. Its civil code Section 207 stated that the ownership of land extends both above and below its surface, subject to limitations by laws and ordinances.

Because the provisions of Section 207 are vague, it can be inferred that many conflicts have arisen among landowners. Some can be seen in legal cases. Some cases followed the judgment noted above, while others restricted land ownership using legal theory such as rights abuse. Although the absoluteness of land ownership has gradually been diluted through these cases, the basic principle established by the two factors exists even today (Kanazawa and Sanbongi 1979; Miyazaki 2006).

On the other hand, for surface water, the River Law (the Old River Law) was enacted in 1896. This opened the door for centralized management. Although the law mostly focused on flood control, some important rules on water utilization, including the permitted water right system, were introduced (the Old River Law § 18).

With economic development, water became important for various purposes such as hydraulic electric power or domestic use in major cities. But the Old River Law did not keep up with these changes. It also lacked provisions to coordinate water uses of different sectors in a river. Reflecting this problem, a series of rules for water conciliation were stipulated, particularly after 1964 when the Old River Law was replaced by the River Law (Study Group on the River Law, Ministry of Construction 1967).

The new rules were aimed at preventing and resolving conflicts over water use between existing water users and new comers. In Japan, surface water is public property and allocation is done by the government (the River Administrator). Stable water supply is essential to society, but river flow varies day by day and year by year. It may be enough to cause flood in a day, but may be scarce enough to cause water shortage in another day. Therefore flow condition must be established first as the basis of allocation.

In Japan, the drought level of water flow (355 days flow volume) in a drought that occurs with probability of once in 10 years is the basic criteria. Within this setting, the first priority is given to the minimum in-stream flow requirements for proper utilization of a river and maintenance of the normal function of it. Next, the second priority is given to existing users. Therefore, it is the remainder that is the subject of allocation for new comers (Okamoto 1995; Murase 2004).

Water allocation is done by government through a permission system. Anyone who wants to use surface water must obtain permission from the relevant River Administrator (the River Law § 23). Even when the new comer is a governmental organization, this rule applies. Because the Kamo River is classified as a Second Class River, the River Administrator is the governor of Ehime prefecture[6] (the River Law § 10). These rules can be interpreted to have a role in providing an advance check on externality problems by restricting free access to rivers.

This permission system leaves room for new applicants in order to promote the effective use of water resources in accordance with changing demands. Let us hypothetically suppose that the Saijo City government has already been a water right holder and that Matsuyama City government tries to take water upstream as a late comer. When the latter diverts water, it could cause harmful effects to existing downstream water users in Saijo City. A series of rules of water use conciliation (the River Law § 38 ~ 43) provides an answer for this problem. These articles require a new applicant to obtain approval from the Concerned River Users who are composed of existing water right holders and fishermen. Basically, the River Administrator cannot give permission without their approval (the River Law § 40).[7] Moreover, if the late comer with permission does harm the existing Concerned River Users, the former must compensate these entities (the River Law § 41). In short, these provisions result in institutional arrangements that foster internalization of the externality by making a new applicant take social costs into account. So if the Saijo City government were a water rights holder for the Kamo River, it would be a stakeholder who enjoys the protection of the River Law.

In the present situation, however, the water users who may suffer from water diversion are not surface water users, but groundwater users.[8] As mentioned previously, the legal status is quite different between surface water and groundwater.

---

[6] By contrast, a river that is of particular importance for land conservation and the national economy is classified as a First Class River. These are under the jurisdiction of the Minister of Land, Infrastructure, Transport and Tourism.

[7] But a River Administrator reserves authority to grant permission to a new applicant without approval of existing Concerned River Users when it is judged that the enterprise of the new applicant will contribute to public welfare more than that of existing Concerned River Users (The River Law § 40).

[8] This does not mean there are none in Saijo City who have water rights for the Kamo River. For example, the Kanbe–Tachibana Land Improvement District has a water right of 1.703 m^3/s (147,139 m^3/day), Ohmachi Land Improvement District 0.992 m^3/s (85,709 m^3/day), Kurare (a company) 1.113 m^3/s (96,163 m^3/day) (Saijo City 2006). Whether Matsuyama City will actually make a formal application for permission for a water right is not yet determined. If Matsuyama City does, it is reasonably to think that it will be necessary for Matsuyama City to obtain approvals from those bodies as a requisite of its permission. But, as mentioned before, this is not an absolute requirement. If the River Administrator concludes the applicant's water use will contribute to public welfare more than that of existing water right holders, it will be able to give permission without approvals of existing water right holders.

The latter is regarded as private water that is subject to land ownership and no permission is required for its use. Therefore, neither the Saijo City government nor individual citizens who use groundwater are included as Concerned River Users. Namely they cannot enjoy the protection given by the River Law.

Thus, the water of the Kamo River, as long as it flows above the surface, is public water subject to the River Law; but once seeping into the underground, it becomes private water and is not subject to this law. Even though surface water and groundwater are often a unit in the physical sense, their legal status depends on where the water is. This boundary brings about a strange situation where the protection of the River Law covers only a user who takes water from above the surface and does not reach to one who takes water from underground.[9] Practically speaking, the Matsuyama City government should pay attention to the assertion of the Saijo City government and its residents, but it cannot be overlooked that these latter users enjoy less protection for their current use compared to the situation if they were to be regarded as concerned River users. What the groundwater problem in Saijo City means is that the legal boundary between surface water and groundwater hinders internalization of an external diseconomy attendant on a water diversion.[10]

15.4 Discussion

Policy implications can be deduced from the groundwater problem in Saijo.

---

[9] Strictly speaking, not all groundwater is regarded as private water. Some kinds of groundwater can be treated as public water as surface water. A typical example is the subflow that runs together with surface water within river banks. Because subflow runs just beneath surface water, diverting subflow may negatively influence the volume of surface water. That is why government permission is necessary for taking such subflow (Study Group on the River Law, Ministry of Construction 1967, Study Group on the River Law 2006; Endo et al. 1975). Some private companies in Saijo City obtained permission from the prefectural government to take water from underground and have been regarded as water right holders for the Kamo River.

There are controversies about how to deal with subflow running out of river banks. This is because the flowing mechanism is uncertain compared to the subflow within river banks. Moreover, it is difficult in practice to cover even this subflow under public regulations (Study Group on the River Law, Ministry of Construction 1967). It is assumed that residents in Saijo City use groundwater that lies in a much deeper area and also runs out of the banks of the Kamo River. Moreover, even though there is no doubt that groundwater is recharged by the Kamo River, the flowing mechanism remains uncertain and residents pump groundwater without government permission. Therefore, we can reasonably postulate that groundwater that is used by residents in Saijo City can be regarded as private water.

[10] Externality between a surface water user and a groundwater user may be mitigated by the in-stream flow requirements mentioned above. The in-stream flow requirement (Kasen-iji-yousui) is defined as "the minimum flow that is necessary for proper utilization of a river and maintenance of the normal function of it". And the flow necessary for maintenance of groundwater level is regarded as one of the examples (Okamoto 1995). It should be noted that Saijo City government requires that the amount of the minimum in-stream flow requirement should be reconsidered in negotiation with Matsuyama City government (Saijo City 2007).

15.4.1 Horizontal Boundary and Vertical Boundary

As noted previously, there are various boundaries related to water that hinder internalization. A typical example can be found in the management of a river that runs across many administrative units such as cities, prefectures, and nations. If each administrative unit has its own River Administrator and they are individually in charge of water rights permissions, they may be able to deal with externalities within each unit but it is difficult for them to deal with trans-boundary externalities. One solution to this problem is the integration of administrative units. If a single authority was responsible for decision-making regarding water rights permission for the whole river basin, the externalities related to the diversion could then be taken into consideration. This solution may be possible in the case of a domestic river but would be difficult when dealing with an international river. Studies have shown that a national border that divides an international river will often become an obstacle to international cooperation on river management (Berber 1959; Elhance 1999; McCaffrey 2001).

This boundary can be called a horizontal boundary in the sense that it divides a natural resource unit and man-made administrative units above the ground. In contrast, the present paper is concerned with a boundary that is drawn vertically; that is, a boundary of differing legal statutes regarding surface water and groundwater. The Kamo River is not fragmentally managed in a horizontal phase: the whole basin is included in the administrating Ehime Prefecture. Nonetheless, the Saijo case shows that inefficient use of water may be brought about as a result of the legal boundary between surface water and groundwater. Even if the gap in the horizontal plane could be resolved, the unresolved boundary regarding the vertical plane will make it difficult to promote the efficient use of water resources. The lesson that can be learnt from the Saijo case is that water management policies need to focus on both the horizontal plane, for example, a river basin, and the vertical plane, the intersection of surface water and groundwater.

15.4.2 Boundaries and Stakeholders

The boundary problem should be considered with special reference to stakeholders. The first step of internalization is to clarify who the stakeholders are. As already mentioned, the system of water use conciliation in the River Law does not work well to prevent externalities between surface water and groundwater users because there is a legal boundary between them. This occurs because the Matsuyama City Government has no obligation to compensate any loss incurred by the groundwater users in Saijo, as long as the latter are not regarded as existing water right holders. In the Saijo case there are two classes of stakeholders. The first are "true" stakeholders based on a natural hydrological unit. The second are the "institutional" stakeholders filtered by the rules of the River Law. Needless to say, only the latter group is entitled to be "stakeholders" in actual institutional settings. So the legal boundary between surface water and groundwater divides stakeholders into two groups: those who can sit at the negotiation table and those who cannot.

The system of water use conciliation is not the only method of promoting internalization. Conversely, if a groundwater user is allowed to ask the Matsuyama City Government to divert less along with some payment, this will also foster internalization.[11] However, this is not likely to be realized as it may be difficult for an individual user to afford the cost. This problem may be resolved if the users work together to make a collective payment. Because groundwater is a shared resource, a lesser diversion will produce common benefits, but it would be fallacious to say that individuals who share a common benefit automatically promote it. When an individual contributes to the common benefit, they receive only a small portion of the benefit their contribution makes. The rest will spill over to other individuals because the benefit is common. If they find the benefit is insufficient and the contribution does not give a return, they will not make any contribution. Rather, they have an incentive to become a free-rider who just expects contributions from others. Where the free-rider prevails, a common benefit will not be sufficiently provided. This is known as the collective action problem (Olson 1965; Dawes 1975). Such difficulties would be avoidable if groundwater users were considered as water right holders.

15.4.3 Justification for Integrated Water Resources Management

The necessity for the integrated management of surface water and groundwater is often advocated in the context of Integrated water resources management (IWRM). IWRM is a conceptual framework of which the definition still remains vague. The meaning depends on the elements that should be considered together in water resources management. The objectives to be considered are: (1) surface water and groundwater; (2) water quantity and quality; (3) water resources and other resources (especially land); (4) water policy and other sector policies (agricultural and energy policies); and (5) stakeholders' opinions (Global Water Partnership 2000; Mitchell 1990; Grigg 1999; Babel 2005; Mitchell 2005).

Among these objectives, integration of surface water and groundwater control is a basic requirement. However, arguments regarding the need for integrated management of surface water and groundwater lack clarity. Arguments for integrated management were made even before the concept of IWRM was developed. For

[11] If groundwater users in Saijo were regarded as existing water right holders, the Matsuyama City Government would need to compensate them. Because of these increased costs, Matsuyama would possibly have to divert less water. Internalization would take place in both situations but the distribution of wealth would differ (Coase 1960).

example, as early as 1929, Wiel insisted that water law should be established and based on geological facts. He maintained the necessity of integrated management from a geological point of view, arguing that surface water and groundwater are often connected (Wiel 1929). This argument is similar to ones made by hydrologists such as Piper and Thomas, who noted that surface water and groundwater are one and the same within the hydrologic cycle and also pointed out that legislation or judicial decisions that were not based on this natural fact resulted in confusion (Piper and Thomas 1958; Thomas and Leopold 1964). Moreover, although perhaps a less persuasive opinion, Dellapenna advocated for integrated management on the grounds that people usually pay attention to the stability of a water supply, not to the source of the water supply (Dellapenna 2003).

As these studies show, earlier researchers have generally indicated the necessity for integrated management based on the physical unity of surface water and groundwater. Although parts of these arguments are persuasive, this is insufficient because they are *abstract* only. Indeed, surface water and groundwater may be a unit in a physical sense but there are also many distinctions in terms of, for example, temperature, quality, and flow speed. It does not follow, just because surface water and groundwater connect with each other, that integrated management is necessary. If this assertion was reasonable, the assertion would also be reasonable that divided management is necessary because of differences in flow speed.

In contrast, the Saijo case illustrates another justification for the integrated management of surface water and groundwater. It shows that a legal boundary between surface water and groundwater will hinder the internalization of externalities. Indeed, it shows that integrated management is necessary, not because both are physical units, but because it will be helpful in preventing wasteful uses of water resources that may be exacerbated by artificial boundaries. This is not the only rationale and another viewpoint could offer an opposite rationale, one against integrated management. We need to compare the real merits and shortcomings of integrated management to resolve this kind of controversy. This paper presents a first step towards that.

15.5 Conclusion

The concept of the hydrologic cycle has significance not only for natural sciences but also for social sciences. As a result of this cycle, water diversion upstream may place external diseconomies on water uses downstream. Thus, conflicts of interests can take place between these uses. A well-designed institution for water resource management would function to prevent or resolve such externality problems. This paper focuses on the influence of human-made boundaries affecting this issue. The investigation is based on the case study of the groundwater problem in Saijo City, Ehime, Japan.

In Japan, a series of rules known as "water use conciliation" are embedded in the national law (the River Law) and have a role in mitigating any externality problems attendant with water diversions. As it is designed to resolve problems among surface water users, it will not function where externality takes place among surface water and groundwater users. This is because there is a legal boundary between surface water and groundwater. While the former is regarded as "public water" and subject to the River Law, the latter is treated as "private water" and belongs to land ownership. So water in a river is public as long as it flows above the surface; but once seeping into the underground, it becomes private water and is not subject to the River Law. Thus, groundwater is out of the jurisdiction of the River Law and a groundwater user is not considered a water right holder whose loss needs to be compensated by the diverter. This means the legal boundary, which does not exist in the natural hydrological cycle, will hinder internalization of the external diseconomy attendant on a water diversion. It has often been asserted, rather vaguely, that integrated management of surface water and groundwater is necessary in the context of IWRM. This paper shows that the efficient use of water resources can be a basis for this assertion.

Acknowledgements This work was financially supported by the Project "Humans and Water" (2005–2009) organized by the National Institutes for the Humanities and KAKENHI (Grants-in-Aid for Scientific Research, Grant-in-Aid for Young Scientists (B)) of Japan Society for the Promotion of Science (Project No. 22710050). Special thanks are extended to Saijo City government for providing materials on groundwater. The views presented here are those of the author and should in no way be attributed to Saijo City government. Responsibility for the text (with any surviving errors) rests entirely with the author.

References

Babel MS (2005) Integrated water resources management (IWRM). In: Lehr L (ed) Water encyclopedia. Wiley, Hoboken, pp 574–576

Berber FJ (1959) Rivers in international law (trans: Batstone RK). Stevens & Sons, London

Coase RH (1960) The problem of social cost. J Law Econ 3:1–44

Dawes R (1975) Formal models of dilemmas in social decision making. In: Kaplan MF, Schwartz S (eds) Human judgement and decision processes. Academic, New York, pp 87–107

Dellapenna JW (2003) The physical and social bases of quantitative groundwater law. In: Beck RE (ed) Waters and water rights, 1991 Edition, 2003 replacement volume, vol 3. Matthew Bender, San Francisco, pp 1–72

Elhance AP (1999) Hydropolitics in the 3rd world—conflict and cooperation in international river basins. United States Institute of Peace, Washington

Endo H, Okawa I, Kanazawa Y, Shiono H, Takahashi Y (1975) On groundwater law. Jurist 582: 16–42 (in Japanese)

Global Water Partnership (2000) Integrated water resources management. TAC background papers no.4

Grigg NS (1999) Integrated water resources management: who should lead, who should pay. J Am Water Resour Assoc 25:527–534

Ichimura M (1909) Public water and private water. Kyoto Hougakkai Zashi 4:54–79 (in Japanese)

Kanazawa Y, Sanbongi K (1979) Water law. Kyoritsu Shuppan, Tokyo (in Japanese)

Mankiew NG (1998) Principles of economics. Dryden, Fort Worth

McCaffrey SC (2001) The law of international watercourses, non-navigational uses. Oxford University Press, New York
Miki A (1993) A history of Uchinuki. History of Saijo 30:11–33
Ministry of Land, Infrastructure, Transport and Tourism (2008) Water resources division. Water resources in Japan
Mitchell B (1990) Integrated water management. Belhaven, New York
Mitchell B (2005) Integrated water resource management, institutional arrangements, and land-use planning. Environ Plann 37:1335–1352
Miyazaki A (2006) Landownership and groundwater law with special focus on the legal characteristics of groundwater. In: Urban and land use (Essays in Honor of Professor Inamoto Yonosuke). Nihon Hyoronsha, Tokyo, pp 47–73 (in Japanese)
Murase K (2004) A review of water rights in Japan. Technical Note of National Institute for Land and Infrastructure Management 156:91–96
Okamoto M (1995) Drought conciliation and water right. Jurist 1060:55–59 (in Japanese)
Olson M (1965) The logic of collective action, public goods and the theory of groups, sixteenth printings. Harvard University Press, Cambridge
Piper AM, Thomas HE (1958) Hydrology and water law: what is their future common ground? In: Water resources and the law. University of Michigan Law School, Ann Arbor, pp 7–24
Saijo City (2000) The result of investigation on Saijo groundwater resource 1996–1999 (in Japanese)
Saijo City (2001) Natural treasure, together with Uchinuki (in Japanese)
Saijo City (2003) Sixty years history of Saijo (in Japanese)
Saijo City (2006) Bulletin Saijo (in Japanese)
Saijo City (2007) A document on the 3rd meeting for opinion exchange on Matsuyama diversion plan (in Japanese)
Saijo City (2010) Annual report on groundwater (in Japanese)
Scitovsky T (1954) Two concepts of external economy. J Polit Econ 62:143–151
Study Group on the River Law (2006) Commentary on the river law, revised edition. Taisei Shuppan, Tokyo (in Japanese)
Study Group on the River Law, Ministry of Construction (1967) Commentary on the river law. Minato Shuppansha, Tokyo (in Japanese)
Suzuki K, Watanabe M, Kadota A (1998) Area of confined groundwater with piezometric head higher than ground level in Saijo city. Mem Facul Eng Ehime Univ 17:209–217 (in Japanese)
Tanaka M (1968) Water right (2). Kensetsu Geppo 21(10):50–53 (in Japanese)
Thomas HE, Leopold LB (1964) Groundwater in North America. Science 143:1001–1006
Wiel SC (1929) Need for unified law for surface and underground water. South Calif Law Rev 2:358–369

Part V
Challenge for New Management Beyond the Boundaries

Chapter 16
Introduction

Osamu Ieda

In Part V we discuss the challenges for new management beyond the boundaries, having four papers concerning various types of transboundary environmental issues: "HELCOM Baltic Sea Action Plan—Ecosystem based approach to manage a semi-enclosed European sea area with nine riparian countries" presented by Juka-Markku Leppanen (Chap. 17); "Combining activities of Sato-umi and Sato-yama in Japan: towards a new type of integrated coastal and watershed management" by Osamu Matsuda (Chap. 18); "Aquifers know no boundaries... but farmers do! So, who should care?!" by Shammy Puri (Chap. 19); and "The transboundary management of groundwater resources in the Kumamoto area; Japan—Sustainable management of groundwater resources for over 700,000 residents"—by Jun Shimada (Chap. 20).

Considering natural resource management systems at diverse scales as shown in the four cases, we had better, first of all, classify the issues by the scale of management, primarily, by the spatial scale of the management. The issues of the Baltic Sea Action Plan and the Middle East aquifers involve international considerations that cross the state borders; the Sato-umi and Sato-yama activities may spread over the prefectural borders; and the groundwater in Kumamoto is an issue involving several local governments. Depending on the nature of the boundaries, each spatial scale requires different types of platforms to negotiate and to manage the issues among the stakeholders. Therefore we shall group the scales of management into four categories: firstly, a local scale for the smallest platform; secondly, a national one for the inter-prefectural or national platform; thirdly, a regional one for the transnational platform; and, fourthly, a continental one for the global platform. No global issues are discussed in the session, though we mention it in the matrix in order to meet the theoretical requirement, since we have to examine global issues as a type of transboundary environmental issue.

O. Ieda (✉)
The Slavic Research Center, Hokkaido University, Sapporo, Hokkaido, Japan
e-mail: ieda@slav.hokudai.ac.jp

M. Taniguchi and T. Shiraiwa (eds.), *The Dilemma of Boundaries: Toward a New Concept of Catchment*, Global Environmental Studies,
DOI 10.1007/978-4-431-54035-9_16,

The other scalar consideration of environmental management relates to time. The papers of the section illustrate various durations over which recovery from the damages on the natural environment could occur. The Kumamoto case, on the one end of the time scale spectrum, shows short term management or action plans to realize the objective. The Middle East case, on the other hand, demonstrates a long term management perspective, which may require even a century-long scale agreement among the concerned countries and peoples. Further we can suppose a much longer scale of duration for environmental management, such as millennium-long, since, for example, an unknown contamination might happen, or might have already happened, within an aquifer lying several thousand meters below the ground, whose discharge usually needs thousands of years.

The chart, "Scale and Scheme Management", proposes a policy matrix, combining the spatial and time scales of the schemes to manage environmental issues.

Scale and Scheme of Management

← Spatial scale of catchment →

Time scale of recovery	Local	National	Regional	Continental
Month	Co	E	E	E
Year	E	L=G	E	E
Century	L=G	P=S	P=S	Cu
Millennium	P=S	Cu	Cu	Cu

List of possible schemes on the Matrix for solution of the environmental issues

Co: Communal scheme: specific solution based on a voluntary agreement

E: Economic scheme: general solution based on a material incentive

L=G: Legal=Governmental scheme: compulsory or normative solution through a local or central authority

P=S: Political=Scientific scheme: political solution based on new scientific knowledge

Cu: Cultural scheme: solution based on a philosophical and perceptional change

We can propose five types of management schemes for environmental issues; Communal scheme, Economic scheme, Legal–Governmental scheme, Political–Scientific scheme, and Cultural scheme. The communal scheme relates to such a way of management which is mostly influenced by the face-to-face relationship among the stakeholders. This scheme aims for an individual solution through voluntary agreement among the stakeholders. This approach is most effective for cases involving local resources and local use. The economic scheme depends on the function of the market or on material incentives. This scheme can function at various scales in space, but is not able to change fundamentally the attitudes of people toward the environment. The legal–governmental scheme is based on a compulsory or normative instrument, guiding people to change their behaviors toward the environment. The case of Sato-yama Sato-umi, though it is not a national issue, shows the necessity of national legal regulation as a basic framework, though it is not always successful in implementation due to, for example, the bureaucratic sectionalism pointed to in the case. Therefore, additional schemes, such as those involving face-to-face relationships or material incentives, may help complement legal–governmental schemes. At any rate, a legal–governmental scheme can be effective at local or national scales; beyond the state borders a legal–governmental scheme can function only as a voluntary norm in the international society. The political–scientific scheme depends on the acceptance of a new scientific knowledge that operates beyond state borders. But a new scientific knowledge is not enough to work by itself as an effective scheme. The knowledge must be implemented by a political leadership committed to creating a new platform to discuss and solve environmental issues. Both cases at the transnational scale exemplify the necessity of collaboration between specialists and policy makers, and also show the serious tensions between these interests. The cultural scheme results from a change in the philosophy or the view of life and death among people. This scheme may amount to a new civilization in regards to the management of the environmental issues.

Allocation of the five schemes in each box of the matrix is very hypothetical and representative. Further, it is important to understand the matrix in such a way that the schemes available at smaller scales may also work in the management of other issues. Namely, for instance, an issue at the regional-decade scale requires an Economic scheme according to the matrix, though we may also expect a legal–governmental or communal scheme to contribute to solving the issue. Among our cases the Sato-yama Sato-umi issue, which is a case of the national-decade scale, successfully introduced an economic scheme, a local money system—a kind of material incentive—at the Fushino River model in Yamaguchi.

The two transnational cases, that is, the Baltic Sea case and the Middle East case are worthy of comparison. Both cases clearly show the necessity of a new scientific knowledge for creating a platform for negotiation among the stakeholders, and the third case, the Amur-Okhotsk case, which is the main topic in Part III of the book, is another good example of the initiative of scientists in a regional transboundary environmental issue. However, the Middle East case demonstrates well that a new scientific knowledge is not sufficient for realizing an international agreement among the stakeholders. On the other hand, the Baltic Sea case demonstrates a successful international agreement, building on the background of European integration. We may say that without regional integration, a transnational environmental management is difficult to realize, since there are no compulsory instruments in the transnational platform. Here, management should involve negotiation among the countries on broader issues in addition to environmental management. Such approach would, eventually, involve material incentives, which shall be a part of transboundary regional development policy, and which could motivate the stakeholders to take part in the negotiation in the initial stages.

As far as the flowchart of each environmental management is concerned, we suppose several common steps. The first step is visualization of the 'unseen' ecosystem. In Chap. 20 the discovery of the ecosystem linkage between the decreasing level of the aquifer and the diminishing rice paddy fields was a first critical moment in the visualization of the 'unseen' ecosystem, and it became an important step in creating an effective policy. Chapter 19 fully discusses the significance of this process, and scientists can greatly contribute to this stage of environmental management. The modeling of the Sato-yama and Sato-umi concepts is also a kind of visualization, though it involves not a new knowledge but a renewal of the traditional way of life in Japan.

The second step is a social recognition of the visualized ecosystem. The particular of appearance of the ecosystem depends on the scale of each scheme as classified in the matrix. The Kumamoto case illustrates an initiative of the local government, and the Sato-yama and Sato-umi case points out the importance of NPO/NGO roles beside and beyond national legislation. Both cases, however, suggest that face-to-face negotiation and mass education campaigns are critical for garnering the support of stakeholders, particularly local residents. The other two cases, concerning the regional scale, tell us that the scientific discourse is crucial. This was all the more true during the Cold War era, when the world was politically divided, and some of the transboundary environmental issues addressed in this book bear this legacy. Science is an effective 'political weapon' in the appearance of environmental issues across state boundaries.

The third step is implementation of the scheme. In this step, motivation of the stakeholders and concerned peoples is ultimately critical, and the scheme option in the matrix is essentially based on the type of motivation required to implement the scheme. The basic dilemma in this step is a gap between the objective and the method. An economic scheme is more acceptable and effective for short term implementation of the policy than the other schemes. However, the motivation in the economic scheme does not come from a consideration for environmental conservation itself, but from a calculation of costs and benefits. In the case of Kumamoto, for example, it would be difficult for the aging farmers or the landowners to reserve paddy fields for environmental conservation in the face of land development pressures in the future. At the time scale of a century, particularly, other ideas should be introduced, such as 'local civil trust' or 'national trust' in a nation-wide issue, which would be required to transform wasted paddy fields to permanent wetlands to preserve water for a sustainable aquifer. This is a transition of the schemes from the communal/economic one to a political–scientific one. In the latter scheme the motivation is not a short run calculation of costs and benefits but an environmentally oriented commitment (a sort of environmentalism).

The philosophical motivation for the cultural scheme is idealistic, though anyone who seriously considers the future of mankind and the global environment realizes that the mass production and mass consumption of modern civilization is not sustainable in the global scale as it has been.

In conclusion, the environment is, by its nature, transboundary. As described in Chap. 19, "Aquifers know no boundaries." Even in a case when an ecosystem is located within a single administrative/political domain, the location was given only historically. Any ecosystem is not closed in time or in space, but is in its nature an open system. Sato-umi and Sato-yama ecosystems have no clear boundaries against surrounding areas. An ecosystem is a convenient or practical discourse and hypothesis when we have to visualize and conceptualize natural phenomena. The basic problem is our way of thinking about boundaries; as again illustrated in Chap. 19, our scientific and management frameworks come from "a boundary in the mind". Transboundary issues in environmental management are visualized lessons which allow us to learn what the environmental issue is, and how to collaborate with people across the boundary based on new ways of thinking. An idea in the mind is realized in the future. Therefore, to change the mind is the way to conserve the environment.

Chapter 17
The Helsinki Convention: 35 Years and Three Eras in Bridging Boundaries to Restore the Marine Environment of the Baltic Sea

Hermanni Backer and Juha-Markku Leppänen

Abstract By definition, environmental issues defy boundaries. This is also true in the northern European Baltic Sea region where the coastal countries have since 1974 cooperated to implement the *Convention on the Protection of the Marine Environment of the Baltic Sea Area* within its governing body—HELCOM. Over the last decades this cooperation has constantly evolved according to changes in political, economic and other types of boundaries. Initial emphasis of the cooperation was to overcome the east–west divide of cold war times by e.g. increasing scientific knowledge on pressures to, and on the state of, the marine environment. However after political changes of the 1990s the focus turned to work to identify, and address, the main pollution sources. The ecosystem approach adopted in 2003, and further defined in the Baltic Sea Action Plan of 2007, called for defining boundaries and limits of a healthy ecosystem as well as further intensified calls for crossing the border between environmental and socio-economic governance for more effective implementation of ecological goals.

Keywords Ecosystem based approach • HELCOM • History • Marine spatial planning • Regional seas • Sustainability • Targets

H. Backer (✉)
Helsinki Commission, Katajanokanlaituri 6, Helsinki 00160, Finland
e-mail: hermanni.backer@helcom.fi

J.-M. Leppänen
Finnish Environment Institute, Marine Research Centre,
PO Box 140, Helsinki 00251, Finland
e-mail: juha-markku.leppanen@ympäristo.fi

M. Taniguchi and T. Shiraiwa (eds.), *The Dilemma of Boundaries: Toward a New Concept of Catchment*, Global Environmental Studies,
DOI 10.1007/978-4-431-54035-9_17,

17.1 Introduction

Environment is a prime example of a topic, which—from the outset—challenges the classical concept of state sovereignty and the hermetic nature of state boundaries. It is in this way clearly an important aspect of the process that is commonly called globalisation. What a state does in, and with, its environment affects not only itself or its immediate neighbours but also the whole regional and eventually global environment. This implies a common responsibility. Such understanding, in addition to other relevant issues, has lead to the conclusion and implementation of numerous regional and global environmental agreements.

The forms of cooperation between participating states following the adoption of such an agreement are, however, not static. In contrary, implementation of environmental conventions often has a dynamic character. Successive amendments, protocols and affiliated declarations resulting from regular Contracting Party meetings create, together with the initial treaty itself, evolving processes often called environmental regimes (Birnie and Boyle 2002). This evolution allows for changing forms and priorities to implementation, guided by, for example, expanding scientific knowledge (Andresen et al. 2000).

One evident source, or facilitator, of such dynamics in regulatory priorities is the evolving geopolitical reality surrounding environmental resources. Boundaries—political, cultural, economic or of other type—between participating states may change or even dissolve altogether, bringing new opportunities for the cooperation.

Such regime evolution and the associated dynamics of boundaries are also visible in the regional work for the Baltic Sea environment carried out to implement the *Convention on the Protection of the Marine Environment of the Baltic Sea Area* (Helsinki Convention) by its governing body, the Helsinki Commission (HELCOM), established with the Convention. Throughout the more than 35 years of HELCOM's history, various major geopolitical, political and economical changes have taken place. Despite such turbulence the organisation has offered a stable platform for dialogue between the Contractive Parties to the Convention.

This paper aims to describe three general phases of the 35 years of evolving regional cooperation for a healthier Baltic Sea carried out within the Helsinki Convention and HELCOM.

17.2 Baltic Sea and Helsinki Convention

The northern European, semi-enclosed brackish Baltic Sea is presently bordered by Denmark, Estonia, Finland, Germany, Latvia, Lithuania, Poland, Sweden and Russia. It covers 337,000 km^2, while its catchment area—1.7 million km^2—extends over an area about four times as large as the sea itself (Fig. 17.1).

The Helsinki Convention has its origins in the fundamental political acceptance that pollution within the Baltic Sea did not respect political boundaries. The scientific evidence and concern about the deterioration of the Baltic Sea ecosystem

Fig. 17.1 The Baltic Sea and its catchment area. *G.o.Finland*: Gulf of Finland, *G.o.R.*: Gulf of Riga, *K.-G.*: Kattegat

in the 1960s[1] and the United Nations Conference on the Human Environment in 1972 paved the way to the first intergovernmental expert meeting about the Baltic Sea in 1973.

As a result of these discussions the Convention on the Protection of the Marine Environment of the Baltic Sea Area was adopted in 1974. A permanent governing body, HELCOM, was established to oversee the implementation of the Convention in 1980.

[1] Emerging from earliest forms of regional cooperation: scientific work carried out within a number of organizations and the academia.

The 1974 Convention, later substantially amended in the form of a new convention in 1992, was at the time a globally pioneering agreement as all relevant sources of pollution were made subject to a single convention. It has as its aim to "... prevent and eliminate pollution in order to promote the ecological restoration of the Baltic Sea Area and the preservation of its ecological balance" (HELCOM 1992). Efforts towards more precise definition of the Convention goals were explicitly initiated with the 2003 adoption of the Ecosystem Approach as a new guiding principle (HELCOM 2003a).

17.3 A Bridge Between East and West—The First Era of HELCOM

A major political boundary in the 1970s was the Iron Curtain between the western and eastern sides of the catchment. This political reality was the clearly the central tone-giving aspect of the first decades of regional cooperation for a cleaner Baltic Sea.

While the Convention made internationally legally binding obligations to be undertaken by each Contracting State, its first immediate effect was to create the legal basis for a close and permanent cooperation among the riparian countries—in the form of HELCOM. The joint efforts in the field of environment worked also as a means to build up overall stability and mutual trust in the region divided by the strong political boundary.

Linked to this political boundary there was also a line drawn in water, separating open and coastal seas. The area for which the Baltic coastal countries were committed to implement actions to fulfil the provisions of the Convention included the whole catchment area of the Baltic Sea. However, the actual area to be protected and restored was limited to offshore sea areas. Territorial waters were left outside due to national security concerns of the Soviet Union.

This was a contributing factor to the focus, and build up of capacity, on issues of maritime traffic but also induced a long-standing open sea bias in HELCOM monitoring and assessment activities. It should be noted that maritime traffic is as a topic excluded from many other regional treaties on the marine environment.

Major achievements during the first decades of the HELCOM cooperation include:

1. An increased scientific knowledge of the state of the entire Baltic marine environment and factors affecting it. This was obtained through harmonised methods used by all countries in collecting data on pollution loads and in monitoring the state of the whole Baltic Sea marine environment. Since then this has made up the foundation of HELCOM's work, making possible production of jointly agreed assessments of environmental status (HELCOM 1981, 1987a, 1993a, 1996, 2002, 2003b, 2009a, 2009b) and compilations of waterborne and airborne pollution loads (HELCOM 1987b, 1989, 1990, 1991, 1993b, 1997, 1998, 2004a), serving as supportive information to decision-makers.

2. Defining "pollution" and adopting a list of specific substances recognised as harmful to marine ecosystems for which HELCOM made Recommendations on elimination (e.g. DDT) or on emission limitations.
3. Establishing a Baltic specific regime to ensure the prevention of pollution from ships navigating in the area and the co-operation in case of major accidents (e.g. involving oil) at sea. Due to the international nature of shipping this required both working for the adoption of appropriate international requirements (e.g. within the International Maritime Organisation, IMO) and their effective and harmonised implementation by the Baltic coastal countries.

The exclusion of coastal waters from the joint regional monitoring and assessment activities gave no possibility to international control. In retrospect, it is clear the information reported to HELCOM by the eastern block did not always match reality as calculated by other means (HELCOM 2004b).

17.4 1992–2003 European Expansion and Civil Society

Political changes in the 1990s dissolved several boundaries in the Baltic Sea area, including the east–west divide. Further, the collapse of the Soviet Union made the enlargement of the European Union in the region possible.

In light of these political changes in Europe, as well as developments in international environmental (e.g. UNCED and affiliated conventions 1992) and maritime law (e.g. UNCLOS 1982), a new updated Convention was signed in 1992 by all the states bordering the Baltic Sea, and by the European Union. In terms of substance the revised Convention was expanded and bridged some sectoral boundaries as it covers, for example, the environmental effects of fisheries and biodiversity in addition to the previous focus on "pollution". The inclusion of NGOs as HELCOM observers signified the opening of this forum of regional cooperation to civil society.

Fundamental new principles and approaches characteristic of this second era included:

1. Inclusion of internal waters of the Baltic Sea to the Convention.
2. A 50% reduction goal had been set for nutrients and a number of hazardous substances in 1988. Linked to this commitment there was recognition of the need to mobilise extra-budgetary financial resources. This included non-national sources and highlighted the consequent need for co-operation and co-ordination between HELCOM and the International Financial Institutions (IFIs). A new approach was used in the work to implement the 1988 Declaration: the definition of "Pollution Hot Spots" which were identified as responsible for a major part of the pollution to the Baltic Sea. The involvement of IFIs ensured that this identification would lead to the designing of projects capable of attracting funding. This work included not only Baltic coastal countries but also other countries in the Baltic Sea catchment area, i.e. Belarus, Ukraine and the Czech and Slovak Republics.
3. An extension of the work of HELCOM to cover nature conservation and biodiversity issues, and by that, the sustainable use of the natural resources

of the Baltic Sea area—including fish and the environmental effects of fisheries.
4. A shift towards a sector-wise approach, including addressing land-based pollution from both point and diffuse sources. This included promoting a Precautionary Approach, Best Available Techniques and Best Environmental Practices rather than setting limit values. In subsequent HELCOM's work on land-based pollution sources, much focus has been placed on the harmonisation of work with similar measures taken within the EU and the sister organisation of HELCOM in the North-East Atlantic, OSPAR.
5. Increased transparency by emphasising public access to information and allowing NGOs to attend HELCOM meetings and work as observers and commentators on the intergovernmental work.

17.5 2003–Ecosystem Approach and Environmental Integration

Even if few political boundaries remained during the turn of the second millennium, the boundaries and limits of the ecosystem, as well as shortcomings in integrating environmental concerns to socioeconomic sectors, ensured that there was no shortage of challenges. The efforts during the 1990s to implement the 1988 Declaration had succeeded in reducing pollution from certain sources like municipal wastewater but major diffuse sources such as agriculture and traffic remained.

In 2003, HELCOM adopted a Ministerial Declaration which reflected EU expansion to encompass by 2004 eight of the nine Member States, as well as further global calls for more action in the field of marine environment and sustainable development[2] (HELCOM 2003a). With the Declaration HELCOM adopted as a guiding principle a new concept, the "Ecosystem Approach to the Management of Human Activities", or in short the "Ecosystem (Based) Approach". The Ecosystem Approach, a concept closely linked to the somewhat forgotten ecological core of the sustainability debate (Bosselmann 2008), had since the late 1990s emerged as an important keyword in national, regional and international environmental commitments (CBD 1998; Rice et al. 2005).

Like its numerous predecessor concepts based on systems thinking (e.g. Waltner-Toews et al. 2008), the Ecosystem Approach emphasises a whole-ecosystem perspective including human societies, as well as science-based cross-sectoral and cross-boundary measures to protect and restore the environment. HELCOM had already incorporated much of such ecosystem-wide considerations inherent to the Ecosystem Approach in the 1974 and particularly the 1992 Conventions. The main novel aspect of the concept included commitments to more quantitative definitions of "good status" of the Baltic marine environment and to using these in defining more precisely required cross-sectoral management actions to reach it.

[2] For example, Johannesburg Earth Summit 2002.

17.6 Quantitative Boundaries of Good Status

In HELCOM, a stepwise approach was adopted to move towards more quantifiable definitions of its aims ("good status" and implicitly the overall goals of the convention itself) by defining in turn a vision, strategic goals, ecological objectives and finally quantitative indicators with explicit target levels (Fig. 17.2).

The vision describes the overall ambition of HELCOM, the strategic goals define major environmental issues of concern and ecological objectives describe central characteristics of a healthy sea (Fig. 17.3). These were generated through a participatory science-policy process and adopted by Member States, to define the topics for which indicators and target levels are to be defined (Backer and Leppänen 2008). Indicators were selected to represent the adopted objectives and

Fig. 17.2 The HELCOM stepwise approach to define good ecological status (modified from Backer and Leppänen 2008)

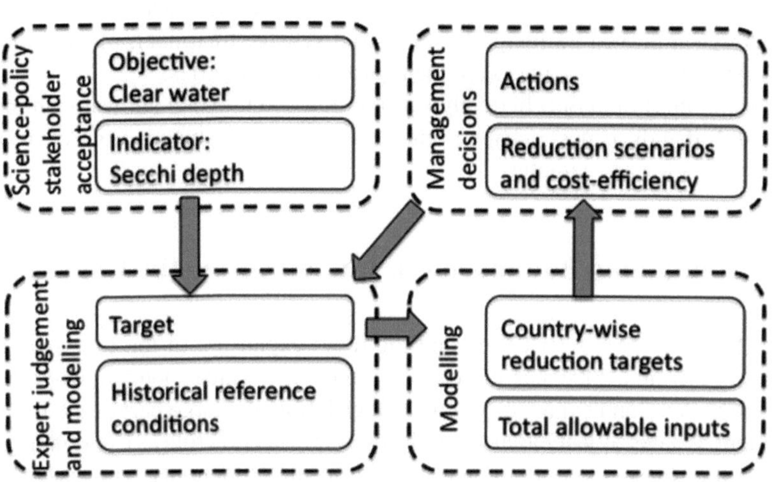

Fig. 17.3 Schematic representation on the linkage between an ecological objective, indicator, target, allowable nutrient inputs and management actions (modified from Backer 2008)

work thus as quantitative proxies of the ecological state (Backer 2008). Targets (usually sub-regionally defined) established the desired indicator value that is the goal of HELCOM.

As an example, the ecological objective "clear water" for eutrophication was translated into a quantitative target of the relevant indicator: summer-time transparency measured with the Secchi disc. Values measured in the first decades of the twentieth century were used as a reference to the pristine status. A pragmatic expert judgement was used to outline acceptable deviation in order to define the sub-regional targets (HELCOM 2006a, 2007).

Management objectives for maritime activities were also developed (HELCOM 2006b) in order to have objectives for all HELCOM main issues.

17.7 From Targets to Actions

Based on the quantitative definitions of ecosystem limits described above, and other relevant work, an intergovernmental programme of measures for the marine environment, HELCOM Baltic Sea Action Plan (BSAP)—explicitly based on the Ecosystem Approach—was adopted in 2007 (HELCOM 2007; Backer et al. 2009). This extensive document includes a number of new issues and approaches but the implications of the quantitative targets, particularly for nutrient pollution, or eutrophication, were the main novelty when compared to previous declarations of this kind.

The Baltic Sea suffers from excessive inputs of Phosphorus and Nitrogen compounds (e.g. HELCOM 2009a). Based on the sub-regional targets for Secchi depth, maximum allowable land-based nutrient inputs were calculated using numerical ecosystem models (Savchuk and Wulff 2007; Wulff et al. 2007). These figures were defined by sub-region and by country and are included in the document as exact numbers.

Management decisions to define the most cost-efficient actions to reach these targets (e.g. improved municipal or industrial sewage treatment, manure handling and other agriculture practices) are then selected by countries.

Nutrient inputs and Secchi depth is monitored and the total allowable inputs will be revised, if judged necessary, to reach the ecological objective of "clear water" (see Fig. 17.3 for a schematic).

17.8 Remaining Boundaries in Implementing the Convention

The collapse of the Soviet Union and the rapid expansion of the European Union between 1990 and 2004 has effectively removed political boundaries within the Convention Area and expanded common legislation to a large part of the Baltic Sea catchment. Conversely, it has also emphasised HELCOMs role as a common ground between the Russian Federation and the European Community. The acceptance of NGOs as observers to HELCOM in the 1990s and their active participation in developing the BSAP adopted in 2007 (Backer et al. 2009) has weakened the boundary between civil society and governments.

However, despite such changes during the HELCOM's lifetime, certain crucial boundaries remain for implementing the convention aims. These include the sectoral boundaries between the environmental regime of HELCOM and relevant socio-economic regimes. The importance of overcoming these cannot be overemphasised.

Achieving sustainable societies, and as a result an environment with its ecological integrity intact, will not ultimately depend on environmental policies but on socio-economic policies guided by sustainability paradigms. Signs of such dissolving sectoral boundaries were seen in the 1992 Convention article on Biodiversity and Nature Conservation, which included the environmental effects of fisheries activities, but these boundaries remain a strong barrier for effective implementation of environmental policy. Similar challenges like those in implementing HELCOM commitments will be faced when implementing the European Marine Strategy Framework Directive regionally (EC 2008).

The example described here of eutrophication in HELCOM BSAP shows how scientific knowledge can be, and has been, used to bridge science and policy by defining more exactly environmental commitments. However, the actual reduction of nutrient inputs from sources like agriculture, which is responsible for the majority of water-borne nutrient inputs entering the Baltic Sea, is related to complex socio-economic regulation. These actions can be decided partly at the national level but the transnational process associated with the EU Common Agricultural Policy largely steering agriculture in the Baltic through subsidies.

Another example of the need for integration across sectoral boundaries in implementation is the challenge to ensure meaningful management of marine protected areas when separate sectoral management initiatives steer fisheries and maritime traffic to the same areas.

Marine Spatial Planning, endorsed by the BSAP and currently developed by HELCOM, coastal states and other actors in the Baltic, is a globally emerging management paradigm where such issues are considered in an integrated way in the spatial context. This approach provides an example of how some of the BSAP actions, and particularly the ecosystem perspective, might be implemented across sectoral boundaries.

Acknowledgements Thanks for Tytti Kontula and Fredrik Wulff for reading early versions of the manuscript. Hermanni Backer has received funding from European Community Agreement with an International Organisation 21.0401/2007/485831/SUB/D2.

References

Andresen S, Skodvin T, Underdal A, Wettestad J (2000) Science and politics in international environmental regimes. Between integrity and involvement. Manchester University Press, Manchester, 221p

Backer H (2008) Indicators and scientific knowledge in regional Baltic Sea environmental policy. ICES J Mar Sci 65(8):1398–1401

Backer H, Leppänen J-M (2008) HELCOM ecological objectives for an ecosystem approach: the process of defining good ecological status of the Baltic Sea. Aquat Conserv: Freshwater Marine 18(3):321–334. doi:10.1002/aqc.851

Backer H, Leppänen J-M, Brusendorff A, Forsius K et al (2009) HELCOM Baltic Sea action plan—a regional programme of measures for the marine environment based on the ecosystem approach. Mar Pollut Bull. doi:10.1016/j.marpolbul.2009.11.016

Birnie P, Boyle A (2002) International law and the environment, 2nd edn. Oxford University Press, New York, 798p

Bosselmann K (2008) The principle of sustainability: transforming law and governance. Ashgate Publ., Aldershot/UK, 242p

CBD (1998) Report of the workshop on the ecosystem approach, Lilongwe, Malawi, 26–28 January 1998, Submission by the Governments of Netherlands and Malawi (UNEP/CBD/COP/4/Inf.9, 20 March 1998) available at www.biodiv.org

EC (2008) Directive 2008/56/EC of the European Parliament and of the Council of 17 June 2008 establishing a framework for community action in the field of marine environmental policy (Marine Strategy Framework Directive). Off J Eur Union L164:19

HELCOM (1981) Assessment of the effects of pollution on the natural resources of the Baltic Sea 1980. Baltic Sea Environment Proceedings 5B. http://www.helcom.fi

HELCOM (1987a) First periodic assessment of the state of the marine environment of the Baltic Sea area 1980–1985; Background document. Baltic Sea Environment Proceedings 17B. http://www.helcom.fi

HELCOM (1987b) First Baltic Sea pollution load compilation. Baltic Sea Environment Proceedings 20. http://www.helcom.fi

HELCOM (1989) Deposition of airborne pollutants to the Baltic Sea area 1983–1985 and 1986. Baltic Sea Environment Proceedings 32. http://www.helcom.fi

HELCOM (1990) Second periodic assessment of the state of the marine environment of the Baltic Sea, 1984–1988; Background document. Baltic Sea Environment Proceedings 35 B. http://www.helcom.fi

HELCOM (1991) Airborne pollution load to the Baltic Sea 1986–1990. Baltic Sea Environment Proceedings 39. http://www.helcom.fi

HELCOM (1992) Convention on the protection of the marine environment of the Baltic Sea Area, 1992 (entered into force on January 17 2000). http://www.helcom.fi

HELCOM (1993a) First assessment of the state of the coastal waters of the Baltic Sea. Baltic Sea Environment Proceedings 54. http://www.helcom.fi

HELCOM (1993b) Second Baltic Sea pollution load compilation. Baltic Sea Environment Proceedings 45. http://www.helcom.fi

HELCOM (1996) Third periodic assessment of the state of the marine environment of the Baltic Sea, 1989–1993; Background document. Baltic Sea Environment Proceedings 64 B. http://www.helcom.fi

HELCOM (1997) Airborne pollution load to the Baltic Sea 1991–1995. Baltic Sea Environment Proceedings 69. http://www.helcom.fi

HELCOM (1998) The Third Baltic Sea Pollution Load Compilation. Baltic Sea Environment Proceedings 70. http://www.helcom.fi

HELCOM (2002) Fourth periodic assessment of the state of the marine environment of the Baltic Sea area, 1994–1998; Background document. Baltic Sea Environment Proceedings 82 B. http://www.helcom.fi

HELCOM (2003a) HELCOM ministerial declaration (HELCOM Bremen Declaration). http://www.helcom.fi

HELCOM (2003b) The Baltic marine environment 1999–2002. Baltic Sea Environment Proceedings 87. http://www.helcom.fi

HELCOM (2004a) The Fourth Baltic Sea Pollution Load Compilation (PLC-4). Baltic Sea Environment Proceedings 93. http://www.helcom.fi

HELCOM (2004b) 30 years of protecting the Baltic Sea-HELCOM 1974–2003. 20p. http://www.helcom.fi

HELCOM (2006a) Development of tools for assessment of eutrophication in the Baltic Sea. Baltic Sea Environment Proceedings 104. http://www.helcom.fi

HELCOM (2006b) HELCOM ecological objectives for an ecosystem approach. HELCOM stakeholder conference on the Baltic Sea Action Plan, Helsinki, Finland, 7 March 2006. http://www.helcom.fi

HELCOM (2007) Baltic Sea Action Plan (agreed 17 November 2007). http://www.helcom.fi

HELCOM (2009a) Eutrophication in the Baltic Sea—an integrated thematic assessment of the effects of nutrient enrichment in the Baltic Sea region. Baltic Sea Environment Proceedings 115B. http://www.helcom.fi

HELCOM (2009b) Biodiversity in the Baltic Sea—an integrated thematic assessment on biodiversity and nature conservation in the Baltic Sea. Baltic Sea Environment Proceedings 116B. http://www.helcom.fi

Rice J, Trujillo V, Jennings S, Hylland K et al (2005) Guidance on the application of the ecosystem approach to management of human activities in the European marine environment. ICES Cooperative Research Report 273:22

Savchuk O, Wulff F (2007) Modeling the Baltic Sea eutrophication in a decision support system. Ambio 36:141–148

Waltner-Toews D, Kay JJ, Lister NE (eds) (2008) Ecosystem approach—complexity, uncertainty and managing for sustainability. Columbia University Press, New York, 383p

Wulff F, Savchuk O, Sokolov A, Humborg C et al (2007) Management options and effects on a marine ecosystem: assessing the future of the Baltic Sea. Ambio 36:243–249

Chapter 18
Combining Activities of *Sato-Umi* and *Sato-Yama* in Japan: Towards a New Type of Integrated Coastal and Watershed Management

Osamu Matsuda

Abstract In the management of coastal marine environments in Japan, the concept of integrated coastal management (ICM) has been officially introduced through establishment of the Basic Act on Ocean Policy in 2007 and the following Basic Plan on Ocean Policy in 2008. These new policies provide for the comprehensive management of watershed and coastal waters beyond the border of administrative sectors. However, implementation of ICM based on the Basic Plan on Ocean Policy has made little progress so far partly due to strong bureaucratic sectionalism. On the other hand, the activities of citizen and NGOs/NPOs which connect the functions of forests, rivers and the sea have recently made significant progress. Among these, combining the activities of *Sato-umi* and *Sato-yama* represents one promising move towards the implementation of ICM. While *Sato-yama* which is widely known as a traditional sustainable land use system in Japan, *Sato-umi* is a relatively new concept which refers to coastal seas managed through sustainable wise use combined with conservation of appropriate natural environment and habitat conditions. Recent local activities connecting *Sato-umi* and *Sato-yama* in the Seto Inland Sea area can be a new type of integrated coastal and watershed management.

Keywords Ecosystem based management • Enclosed coastal sea • Environmental management • Integrated coastal management (ICM) • Pollution load control • Resource management • *Sato-umi* • *Sato-yama* • Sectionalism • Seto Inland Sea

---

O. Matsuda (✉)
Graduate School of Biosphere Sciences, Hiroshima University,
6-8-13 Hachihonmatsu-minami, Higashi-hiroshima 739-0144, Japan
e-mail: matsuda036@go3.enjoy.ne.jp

M. Taniguchi and T. Shiraiwa (eds.), *The Dilemma of Boundaries: Toward a New Concept of Catchment*, Global Environmental Studies,
DOI 10.1007/978-4-431-54035-9_18, © Springer Japan 2012

18.1 Introduction

Many scientific studies have shown that the condition of coastal ecosystems and living marine resources around Japan have deteriorated remarkably. A 2010 assessment of biodiversity in Japan, issued by a panel of expert from the Ministry of the Environment, describes serious losses in the country, particularly in such ecosystems as terrestrial waters, coastal areas and the sea, and islands (Ministry of the Environment 2010).

Although existing mechanisms for linking science and policy are highly sectoral, the major coastal problems are increasingly multisectoral. Integrated coastal management (ICM), which has long been proposed in Japan, includes comprehensive management of watershed and coastal waters beyond the border of administrative sectors relating to the management of forests, rivers, land, agriculture, coastal environments, ports and harbors. The concept of ICM has been officially introduced to the legal system in Japan only recently through establishment of the Basic Act on Ocean Policy (the 25th article) in 2007 and the following Basic Plan on Ocean Policy (Section 9 of Part 2) in 2008. Implementation of ICM based on the Basic Plan on Ocean Policy has so far made little progress, partly due to the strong bureaucratic sectionalism among related Ministries. From the viewpoint of ICM, the Basic Plan on Ocean Policy did not realistically embody the concept because interrelationships among responsible Ministries and sectors remained ramified and complicated. Realistic implementation of ICM by the central government in Japan will take more time.

While ICM has been difficult to implement through official policy, the activities of NGOs/NPOs and local people which integrate the functions of forests, rivers and the sea have made significant progress toward ICM principles. Among these, the recent combination of the practices of *Sato-umi* and *Sato-yama* is a particularly promising movement. *Sato-yama* and *Sato-umi* can be both defined as dynamic management systems that integrate people and nature in traditional agriculture, forestry and fisheries in Japan.

In the following sections, *Sato-yama* and *Sato-umi* are first introduced and then the present status of environmental conditions in the coastal seas in Japan are described, followed by a description of the relationship between *Sato-umi* and national policy. After describing the international response to *Sato-umi*, finally, two case studies in the Seto Inland Sea area, which combine *Sato-yama* and *Sato-umi* towards a new type of integrated coastal and watershed management, are introduced.

18.2 What Is *Sato-Yama*?

Sato-yama is a Japanese term for traditional rural landscapes. In Japanese, *Sato* means local village or community where people live and *Yama* means forest or forested mountain. Historically, people made effective use of forests near rice paddies and agricultural fields. For example, fallen leaves were used as fertilizer, the trees themselves were used as wood or fuel, and people went into the forest to gather edible plants and vegetables. This type of traditional area, where the forest plays a vital role in the livelihood of rural people, is called *Sato-yama* in Japanese. This type of sustainable use of forest was made possible by prolonged interaction between human life and forests, in particular by the regular cutting of broadleaf trees. However, the term Sato-yama refers not only to the forest but also to the variety of landscapes, such as paddy fields, waterways, levees between fields, and nearby reservoirs, which are closely associated with them.

18.3 What Is *Sato-Umi*?

In Japanese, *Umi* is the most common word meaning the sea. The concept of *Sato-umi* originally comes from *Sato-yama*. The basic idea of *Sato-yama* applied to the coastal seas is called *Sato-umi*. The concept of *Sato-umi* as "high productivity and biodiversity in the coastal sea area with human interaction" was first proposal by Yanagi (1998) and developed by Yanagi (2007). However, the term *Sato-umi* is now often used in a broader sense to indicate any appropriate relationship between humans and the coastal seas, including creative activities of local people towards more productive and functional seas. Therefore, the functional characteristics of *Sato-umi* are a coastal sea under the harmonization of sustainable wise use by humans along with the conservation of appropriate natural environment and habitat conditions. *Sato-umi* should provide higher biological diversity as habitat and higher biological productivity as fishing ground (Matsuda 2010). *Sato-umi* is also a suitable site for demonstrating the multi-functional roles of fisheries (Matsuda 2008a).

The combined management of *Sato-yama* on land and *Sato-umi* around coastal sea areas beyond existing boundaries is more effective from the view point of water resources management and land–sea interactions such as nutrient cycling and material flow (Babel 2003). However there are many existing boundaries between *Sato-yama* and *Sato-umi,* including legal, administrative, geographical and traditional resource management boundaries, which prevent promotion and implementation of ICM.

18.4 Environmental Condition of Coastal Seas in Japan

This section provides a general overview of environmental conditions of coastal seas in Japan and then describes the conditions in the Seto Inland Sea is separately described as the Seto Inland Sea as *Sato-umi* since the Seto Inland Sea has long history as *Sato-umi* and is a central part of new activities on *Sato-umi*.

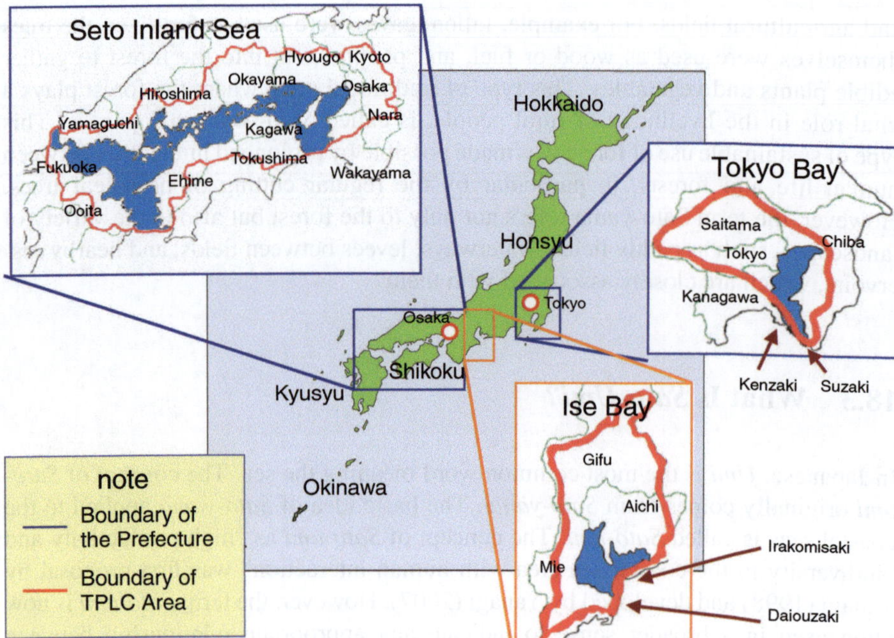

Fig. 18.1 Three TPLC (Total Pollutant Load Control) designated areas and responsible prefectures in Japan (MOE)

18.4.1 General Overview of Coastal Conditions

In order to improve the serious water pollution which occurred during the period of high national economic growth from the mid-1960s to mid-1970s, effluent water quality standards on nitrogen and phosphorus have been applied in enclosed coastal seas in Japan. In particular, in the three enclosed coastal seas of Seto Inland Sea, Tokyo Bay and Ise Bay and their associated watersheds, area-wide Total Pollutant Load Control (TPLC) system based on a Chemical Oxygen Demand (COD), total nitrogen (TN) and total phosphorus (TP), have been applied as a policy of the Ministry of Environment (MOE) (Fig. 18.1). This area-wide TPLC system, which includes both watershed and coastal areas, was a first step connecting watershed and marine environments across coastlines. As a result, serious water pollution in coastal areas was remarkably reduced. However, functional depression of material circulation and deterioration of other ecosystem characteristics including biodiversity, habitat and living resources, are still serious. Lack of concern for marine environmental conditions is a concern. Therefore, in addition to improvement of water quality, restoration of deteriorated biodiversity, habitat, and living resource conditions and active involvement of citizens have become more important. The major objectives of coastal environmental management in Japan have gradually shifted from water quality control through regulation to more holistic restoration of the environment and ecosystem with the participation of a variety of people under the ICM framework, as exemplified by the creation of *Sato-umi*.

18.4.2 The Seto Inland Sea as Sato-Umi

The Seto Inland Sea's long history of providing ecosystem services to humans can be seen as a *Sato-umi*. However, serious water pollution and deterioration of the ecosystem occurred during the postwar reconstruction period after WWII. Rapid economic growth between the mid-1960s and mid-1970s was accompanied by serious water pollution and eutrophication in the Seto Inland Sea (International EMECS Center 2003).

Among the many countermeasures which were undertaken, area-wide TPLC system in terms of COD, TN and TP has played an important role in the improvement of water quality. However, reduced shallow sea area by land reclamation, decreases in numbers of animal and plant species, and deteriorated habitats have still not been recovered (Matsuda 2008b). Therefore, the focus of environmental management of the Seto Inland Sea has recently shifted from water pollution control to wider goals such as the conservation of biodiversity and biological productivity and the restoration of habitat and well-balanced nutrient cycling through such initiatives as the creation of *Sato-umi*.

During the last 40 years, shallow areas, in particular tidal flats and seagrass beds, have been drastically lost in the Seto Inland Sea, mainly due to land reclamation. These shallow areas provide valuable habitat for many organisms. Seagrass beds are known as excellent spawning and nursery grounds for many species and are often called "cradles of the sea". Because the recent decrease in fish catches, particularly of benthic species, is strongly due to the loss of shallow areas and deterioration of sediment quality, environmental restoration of shallow areas is essential for the management of living marine resources and habitat and biodiversity conservation.

Coastal environmental management in Japan, in particular applications of TPLC, have been quite successful at reducing the negative effects of eutrophication. However, such targeted measures as TPLC could not provide for the recovery of deteriorated habitats, ecosystems and living resources. More holistic ecosystem based approaches, such as combining the practice of land management through *Sato-yama* with the practice of coastal management through *Sato-umi,* proved necessary in order to restore the deteriorated coastal environment and ecosystem.

18.5 *Sato-Umi* as a National Policy

The creation of *Sato-umi* as a national policy of Japan was initiated when it was designated as a near term environmental policy priority in the Becoming a Leading Environmental Nation Strategy in the twenty-first Century report (MOE 2007). Through the implementation of this national strategy and related projects, the efforts of local people to create new *Sato-umi* have been supported. Among the many new ICM approaches, the *Sato-umi* initiative is promising because it has been incorporated into national policy. The promotion of *Sato-umi* should provide for the maintenance of high biological diversity and high biological productivity through the collaboration of a wide variety of groups and stakeholders.

Sato-yama and *Sato-umi* related activities in Japan are also promoted through a related national policy, the Ministry of Agriculture, Forestry and Fisheries' Strategy for Biodiversity Conservation, which was established in July 2007. The major emphases of this strategy are that the negative effects of agriculture, forestry and fisheries on biodiversity in the past should be improved and that affected ecosystems should be restored.

Appropriate future initiatives are conservation and restoration of coastal marine environments such as tidal flats and seagrass beds, as well as implementation of ecosystem based management (EBM) approaches for living resources and establishment of appropriate marine protected areas (MPA). Through these approaches, strong linkage among the management of forests, rivers and marine environment beyond existing boundaries is very important.

18.6 International Responses

The oncept of *Sato-umi* received high praise at the conclusive review of the seventh International Conference on Environmental Management of Enclosed Coastal Seas (EMECS7) held in France in 2006 based on the potential of *Sato-umi* to provide for symbiosis among human communities and coastal/marine areas. In 2008, EMECS8 was held in Shanghai, China with the main theme of "Harmonizing River Catchment and Estuary". About 470 participants from 37 countries participated in this meeting. A *Sato-umi* session was held and the "Shanghai Declaration," which was adopted on the last day of the meeting, recognized *Sato-umi* as a promising and constructive new concept for coastal management (International EMECS Center 2009). In the East Asian Seas Congress (EAS-Congress) 2009, which was held in November 2009 in Manila, Philippine, a *Sato-umi* workshop was co-organized by the Partnerships in Environmental Management for the Seas of East Asia (PEMSEA) and the International EMECS Center with the theme of "Indigenous Approaches to Habitat Protection and Restoration: Experiences in *Sato-umi* and other Community Initiatives" (International EMECS Center 2010). This workshop aimed to share information on innovative community-based approaches and effective practices of habitat prevention, restoration and management applying low-cost technologies, to identify and distill lessens from specific case studies, and to examine how such local community initiatives can be further developed, packaged and extended within and across jurisdictional boundaries. This workshop, which was chaired by the present author, included a broader discussion on the *Sato-umi* concept and similar ideas in the context of the ICM framework as well as presentations of case studies on their application in Asian countries.

Another international initiative is the ongoing "Sub-global Assessment of *Satoyama* and *Satoumi* in Japan" (*Satoyama-Satoumi* SGA). This project is a follow-up to the sub-global assessment (SGA) of the Millennium Ecosystem Assessment (MA) proposed by United Nations during 2001 through 2005. In this *Satoyama-Satoumi* SGA, *Sato-yama* and *Sato-umi* in Japan were classified into five clusters from the Hokkaido Cluster to the Western Japan Cluster. The project assesses the interrelationships between and possible integrated management of *Sato-yama* and *Sato-umi* beyond the existing boundaries between *Sato-yama* and *Sato-umi*. The Seto Inland Sea, which is the largest enclosed coastal sea in Japan, is included in the Western Japan Cluster and is playing an important role as a typical example of *Sato-umi*.

18.7 Towards a New Type of Integrated Coastal and Watershed Management

Although implementation of ICM based on the national Basic Plan on Ocean Policy is making little progress partly due to strong bureaucratic sectionalism, there are many examples of activities by local people in which the practices of *Sato-yama* and *Sato-umi* are connected. Two cases of such activities are introduced below.

18.7.1 Case of Fushino River Watershed and Estuary

In the watershed and estuary of the Fushino River in Yamaguchi prefecture, the combination of *Sato-yama* and *Sato-umi* has been supported by a variety of people including fisherman and foresters.

Fushino River is located near the west end of Honshu Island and faces the Suo-nada area of the western part of the Seto Inland Sea. The watershed area of Fushino River is 322 km^2 and the length of the river is about 30 km. The area of the estuary is about 1,700 ha with a tidal flat of about 350 ha. Population in the basin is about 108,000.

People of the basin designed the plan for integrated management of the watershed and the estuary, from the mountain forest to the sea. Some important key phrases in the plan were "local consumption of local products" and "think of the source when you drink water". Basic principles of the project included cooperation of local stakeholders based on scientific knowledge and adaptive implementation of the management program. Administrative aspects of the project are largely handled by the environmental section of Yamaguchi prefecture government.

Major management issues in the estuary included the decrease of fish catches, especially of short-neck clam, decrease in seagrass beds, and the protection of endangered species such as the horseshoe crab. These problems seem to result from human activities around the area, including forestry, agriculture, fisheries, industries, construction and land reclamation, and waste treatment during the last half century.

The main restoration objectives of the *Sato-umi* and *Sato-yama* in the estuary and watershed of Fushino River include restoration of short-neck clams and seagrass beds through various approaches, and surveying the distribution and behavior of horseshoe crabs. Cleaning of beach and river from upstream to downstream, and planting trees in the upstream forest were also conducted with the cooperation of upstream people such as foresters and downstream people including fisherman and local citizens. Local money called "Fushino" was issued to stimulate the involvement of local people to participate in those restoration programs.

As a result of these activities, the area of seagrass bed has been gradually recovering and plowing of tidal flats with net-covering has proved to be effective for promoting clam shell production (Fushino River estuary research working group 2009).

18.7.2 Case of the Akou Coast and Chikusa River Watershed

The activities of local people and children aimed at restoring the seagrass beds of the Akou coast in Hyougo prefecture, which is connected to the conservation of Chikusa River watershed, also reflect the combination of *Sato-yama* and *Sato-umi*.

The Akou Coast and Chikusa River watershed is also one of representative sites for environmental restoration based on the concept of *Sato-umi* and *Sato-yama* in the Seto Inland Sea. Since a variety of activities related to environmental restoration and education have been developed in this area, a variety of stakeholders and local people joined the community-based or co-management activities. Among the variety of activities, typical examples include an activity on the conservation of the Chikusa River watershed initiated by NGO groups and an activity on seagrass bed restoration by voluntary divers. Recently, these activities have been integrated as a part of a *Sato-umi* project supported by the central government MOE, Hyogo Prefecture, and the local government of Akou City.

Major targets of this *Sato-umi* project are restoration of the deteriorated seagrass bed and unique shore vegetation along the Akou coast. Restoration of decreased levels of short-neck clams in the area and promotion of environmental education are also important objectives of the project. Among many related activities, one of the highlights is the achievement of Misaki Elementary School in which school children raised seedlings of eel grass and planted them in the targeted seagrass bed area.

During the process of the *Sato-umi* project, many regional meetings were held in order to share the basic idea of the restoration program based on the concept of *Sato-umi*. On the occasion of the *Sato-umi* Symposium held in Akou in March, 2009, many people from community-based activity groups assembled at the City Hall and discussed future plans. This indicated the strong involvement of a variety of relevant persons and groups in the *Sato-umi* project. The establishment of the Akou Coast *Sato-umi* Committee for co-management in this area is now within reach.

Historically, Akou Coast was well known for the traditional salt making industry and local people made effective use of the coastal area. A part of the old salt pan was already converted to a public Seashore Park by the prefectural government and included the Salt Industry Museum. Since the Seashore Park is one of the activity centres for local people, it is expected that the activities of the *Sato-umi* project will be combined with the Seashore Park activities.

18.7.3 Possible Approach Towards a New Type of ICM

In the case of the Fushino River watershed and estuary, cooperation between upstream people and downstream people has been quite successful and has involved a variety of people and groups.

In the case of the Akou Coast and Chikusa River watershed, cooperation between NGOs and school education programs contributed effectively to environmental restoration. These activities of local people combining *Sato-umi* and *Sato-yama* in the Seto Inland Sea area represent a promising approach towards a new type of integrated coastal and watershed management in Japan.

Apart from the Seto Inland Sea area, there are many case studies demonstrating the combined activities of *Sato-yama* and *Sato-umi* in Japan. These activities were reported as part of the *Satoyama-Satoumi* SGA in the Convention of Biological Diversity (CBD)-COP10 which was in Nagoya, Japan in October, 2010.

References

Babel MS (2003) Integrated water resources management (IWARM). In: Lehr L (ed) Water encyclopedia. Wiley, Hoboken, pp 574–576

Fushino River estuary research working group (2009) Report of the restoration of living resources in an estuarine tidal flat, 444 pp (in Japanese)

International EMECS Center (2003) Seto Inland Sea. In: International EMECS Center (ed) Environmental guidebook on the enclosed coastal sea of the world, pp 65–72

International EMECS Center (2009) Report of International Workshop, Sato-Umi new concept that increases biological productivity and biodiversity, 138p

International EMECS Center (2010) EAS Congress 2009 Sato-umi workshop, workshop report, 154pp

Matsuda O (2008a) *Sato-Umi* as new approach for the restoration of the Seto Inland Sea. Science Council of Japan Forum 6:17–23 (in Japanese)
Matsuda O (2008b) Eutrophication and its causes/consequences: The case of the Seto Inland Sea. In: Asia-Pacific coasts and their management. Springer, Dordrecht, pp 78–92
Matsuda O (2010) Recent attempt towards environmental restoration of enclosed coastal seas: Ago Bay restoration project based on the new concept of *Sato-Umi*. Bull Fish Res Agen No 22:9–18
Ministry of the Environment (2007) Becoming environmental nation strategy in the twenty-first century report. pp 1–27 (in Japanese)
Ministry of the Environment (2010) Japan biodiversity outlook (Brochure in Japanese)
Yanagi T (1998) To create "Sato-Umi" in the coastal sea area. J Water Environ Soc 21:703 (in Japanese)
Yanagi T (2007) Sato-Umi: new concept for coastal sea management. TERRAPUB, Tokyo, 86p

Chapter 19
Aquifers Know No Boundaries… But *Farmers* Do! So, Who Should Care?!

Shammy Puri

Abstract Human endeavours are largely governed by "boundaries", seen and unseen. Our social interactions are subject to boundaries that are set in cultural norms: certain activities are permitted and maybe encouraged, others are accepted but not encouraged, and yet others are completely taboo. Our perception of the world around us is also governed by boundaries though some of these may be very blurred boundaries.

Just as our daily lives are governed by boundaries, so our study of science is full of boundaries, explicit and implicit—for instance, the boundary between the science of biology and physics which was once rigid and is now very blurred. This chapter stems from discussions that took place in the 4th International Symposium hosted by the Research Institute for Humanity and Nature in Kyoto, Japan, on The Dilemma of Boundaries—Towards a New Concept of the Catchment, and particularly from the discussion that related boundaries in aquifers to the boundaries that a farmer might respect in his use of water in aquifers. Reflection on this dilemma suggests that it is us humans that need to make "boundaries" in a natural world, which is in fact a continuum, and exists in a smooth transition, from one state to the next.

In making a study of the Rum-Saq Aquifer system (Puri, Aquifers know no boundaries, Guest Commentary in Journal of International Groundwater Technology, p 6, 1997), the present author stated that, "aquifers know no boundaries" (except hydraulic ones) because it became clear that the area of study that initially was only several hundreds of square kilometres had to be extended to several thousand square kilometres. The experience gained from that study led the author to be a proponent of the study of transboundary aquifers, which has now been recognised through a UN resolution that encourages countries that possess transboundary aquifers to refer in their approach to these aquifers to Draft Articles prepared by the International Law Commission and grounded in legal formulations. This chapter addresses the conceptualisation of boundaries, their characterisation and their calculation.

S. Puri (✉)
Secretary General, International Association of Hydrogeologists,
Dorchester on Thames, Oxon OX10 7PJ, United Kingdom
e-mail: ShammyPuri@aol.com

M. Taniguchi and T. Shiraiwa (eds.), *The Dilemma of Boundaries: Toward a New Concept of Catchment*, Global Environmental Studies,
DOI 10.1007/978-4-431-54035-9_19,

Keywords Boundaries in a globalised world • Boundaries seen and unseen • Implicit and explicit • Jordan and Saudi Arabia • The Rum-Saq aquifer • Transboundary aquifers

19.1 Introduction

The allocation of land and property is governed by boundaries that are commonly defined in legal title claims. Claims to goods or services that appear as common pool resources are of equal validity though they are accessed without the provision of conventional legal documents. For example, a farmer who owns land and has the ultimate right to it, also has rights to sunshine, rain and water which are a common pool resource. This paper discusses the concept of boundaries that separate or divide, which are present in many aspects of natural resource management, and which may result in either cooperative or conflictive situations.

The discussion starts with an overview on the perceptions of boundaries and then considers the recognition of boundaries in the physical world and in science. In trying to understand our world, there is an inherent need to classify, categorise and conceptualise. The discussion also considers boundaries relative to smooth transitions which are usually the more realistic representation of natural phenomenon. At a planetary level, it would seem that boundaries in a globalised world have to be viewed in a different light, taking into account the impact of such issues as climate change and its increasing variability. The impact of globalisation on the environment is a matter of concern, while 'globalisation' has also made the issue of boundaries in international financial transactions more or less defunct. Globalised trade in agricultural goods can impact the fortunes of farmers at the local level through the transfer of virtual water trapped in their products (Allan 2001).

The final section of the discussion will relate to an example concerning the transboundary aquifer that occurs on either of the national boundaries of Jordan and Saudi Arabia. This aquifer is a critical water resource in an arid region with a potentially catastrophic water scarcity problem in the coming decade. Many regions of the world have transboundary aquifers that are an important source of water and lie across national boundaries. The experience of the conditions in the Jordan–Saudi Arabia area has influenced the author to propose international tools that may help overcome the potential problems that boundaries cause for global natural resources management and consequently for the users, such as farmers, of such resources.

19.2 Perception About Boundaries

Practically all aspects of human endeavours are somehow governed by "boundaries", seen and unseen (Blomquist and Ingram 2003). Human social interactions are subject to boundaries that are set in cultural norms: certain activities are permitted and maybe encouraged, others are accepted but not encouraged, and yet others are completely taboo. Each day in our lives we either stay within these boundaries, go close to them or indeed cross them.

Our perception of the world around us is also governed by boundaries, though some of these may be very blurred boundaries. Every farmer explicitly knows the boundaries and the physical properties of his land, but also implicitly knows of the wider boundaries from which water, essential to his very existence, comes. While the farmer has ultimate and total control of his owned land, what control can he exert over the wider implicit boundary of the 'water system' available from common pool resources, over which he may have very little explicit control? (Oel et al. 2007)

19.3 Explicit and Implicit Boundaries

Everyone can tell whether one is in a city or has left it to reach into the countryside—and even in the absence of a roadside sign telling us that we are leaving or entering a city, we can tell whether we are in or out of it, though this will be intuitive, because there is no rigid boundary between town and country.

There are boundaries that occur in nature that we can readily recognise; the point at which land ends and the sea starts is obvious, though the point at which hilly lands start and the plains are behind us is a blurred boundary, though we can nevertheless tell.

There are boundaries that have been drawn up through cultural norms within many societies—however, as this aspect of boundaries requires good knowledge of social sciences, which is beyond the scope of this paper (Table 19.1).

19.4 Classification, Categorisation and Conceptualisation

Just as our daily lives are governed by boundaries, so our study of society, science and nature is replete with boundaries, both explicit and implicit. For example, the boundary between the sciences of biology and chemistry was once rigid and is now very blurred. The study of social sciences and the study of physics would seem to have a clear boundary between them, while the boundary between the study of social sciences and law, or jurisprudence, may be very blurred, or gradual. Are these boundaries real or are they imagined?

Table 19.1 Various boundaries

Explicit boundaries	Implicit boundaries	Intuitive boundaries	Blurred boundaries
'Seen boundaries'	'Unseen' boundaries	'Unseen boundaries'	Adopted boundaries
In nature	In the mind	Cultural and customary	Land use and eco-climatic zoning

The title of this RIHN Symposium correctly encapsulates the problem—the dilemma of boundaries. At first sight one may question whether indeed there is a dilemma, and if so, whether the dilemma matters to us, or if it is only a metaphysical matter? To answer this questions, it is worth considering some occurrences of boundaries in science and in the globalised world, as well as an alternative concept, that of smooth transitions from one condition to another.

19.4.1 Boundaries in Science

While not a comprehensive assessment, the following lists some of the areas where boundaries are invoked for conducting assessments and analysis. In the study of natural phenomenon, boundaries are invoked for good reasons as they permit the isolation of different characteristics of nature. The geological boundaries between rocks of different ages enable a clearer understanding of the conditions in which organisms evolved and adapted to prevailing climatic conditions. The boundaries invoked in hydrology enable an assessment of the sustainability of agriculture and ecosystems to be made, while the boundaries invoked in biology allow a clearer understanding of the ability of organic life to survive. Boundaries are also invoked in the effort to understand human relations, and these include boundaries on the basis of race and religion, as well as creed and culture. While invoking boundaries in the practice of science remains an acceptable and non-controversial issue, the invoking of boundaries in human relations can often be a recipe for possible tensions.

Although ‘multi-disciplinarity’ or boundary spanning (Warner et al. 2010) is often required to resolve many of the more complex problems that are faced by humanity, it nevertheless poses significant difficulties in conceptualisations—experience has shown that even though ecologists and technologists faced with, for example, water scarcity-related investments, will work alongside social scientists who assist with analyses of the water users—farmers or other communities—they nevertheless face communication lapses. Efforts are being made to understand this type of complex relationship, as studies of hydro-solidarity, hydro hegemony (Zeitoun and Warner 2006) and similar studies show that the physical boundary of a watershed and socio-economic inter-relationships cannot be separated, though addressing them is not straightforward.

In a series of recent reports published in the Journal *Nature* (Nature Reports Climate Change: Published online: 23 September 2009) a group of scientists presented a framework on ‘planetary boundaries’ in which they addressed seven global environmental indicators of thresholds. As judged by them, crossing the threshold of even one of these thresholds could lead to the instability of the entire earth system. However, defining an acceptable upper limit, or a tipping point, for any of the indicators has been thought by some to be somewhat arbitrary. Thus in recognising that the earth system can only be defined by an intricate combination of several

Table 19.2 Natural and human boundaries

Study of natural phenomenon	Study of human phenomenon	Multi disciplinarity
Boundaries in Geology	Racial boundaries	Blurred boundaries—overlaps—inter digitations of boundaries—time and space limited boundaries—hydro-solidarity—hydro-hegemony
Boundaries in Hydrology	Religious boundaries	
Boundaries in Biology	Cultural boundaries	
	Linguistic boundaries	

sciences, there is the need to focus on multiple boundaries—firstly those between the sciences and secondly in the definition of upper limit or bounds. As an aside the findings of this study are truly worrying for the fate of the planet and the reader is urged to review it (Table 19.2).

19.4.2 Boundaries in a Globalised World

Following on from the discussion in the previous paragraph, in the course of the past decade, the advent of a new condition, known as globalisation, has captured the thinking of economists, politicians and latterly also scientists (Puri 2006). The term globalisation has taken on many meanings. As an example, the mitigation of planetary climate change, referred to above, requires globally concerted actions, but experience shows that our moral principles of action are only effective at a local level. These local actions seem to be tied closely to culture, custom or religion, and are sometimes in divergence with each other. How then to develop a notion of joint global ecological responsibility? This is the most important question in the debate about climate protection, since *motivation* is the key to individual change and *responsibility across borders* is the key to global change. When applied to water, the issue is more intricate, since the drivers of climate change relate to invisible transmissions through release of CO_2 to the atmosphere, which does not have boundaries, while drivers to water resource change would seem to only be transmitted through some action on-the-ground and within some boundaries (e.g., national). However, as the now fashionable study of the flow of virtual water shows, demand for goods supplied through multinational corporations does influence the water resources of countries that have no common physical border nor common physical watersheds.

There are yet other types of problematic boundaries in a globalised world. Take for instance the boundaries of expected global temperature increases—climatologists have indicated that the boundary of 3°C global temperature rise would lead to serious impacts on the globe. Although the science suggests that up to 6°C rise can be expected in the forthcoming 90 years (2100), already at 3°C rise we can expect rainforest die back and the loss of major rivers.

Table 19.3 Natural abrupt boundary and smooth transitions

Natural abrupt boundary	Natural smooth transitions
Faults in geological rock formations	Lithological change in a sequence
Terrestrial—marine junctions	Species of trees in forests
	Habitats of fauna

19.4.3 Responsibilities Across Borders

The foregoing discussion has pointed out that while we work with all manner of boundaries, in some cases we need to consider responsibility for actions that lie across a boundary. Here we would need to analyse the players, by asking, 'Who are the actors?' While, a response that 'we all are' may be intellectually painless, perhaps it is policy makers who really carry the responsibility. The decisions made by policy makers after all would enable responsible investments to be made, corporate responsibility to be seriously taken, and intercultural ethics to be practiced more effectively for achieving a form of sustainability. Multinational corporations have found that cross-cultural differences in value commitments and models of responsibility matter for inter-cultural marketing of their goods, and they need to adopt to new management and negotiation models to behave responsibly across borders. Once this is in place then it may be possible to assess whether businesses are merely 'green-washing' their interests or aiming for genuinely responsible actions.

19.4.4 Boundaries Versus Smooth Transitions

While the dilemma of boundaries may not have simple answers, I would venture to suggest that it is us humans that need to make "boundaries" in a natural world, which in fact is a continuum, and is characterized by smooth transitions, from one state to the next. The change from flat plains to hilly lands is in fact a smooth transition; the change from one species of flora in a natural landscape to another is also a smooth one. The natural hydrochemistry of water in an aquifer system has a smooth transition. For each of these 'smooth' transitions, for the purposes of conceptualisation, humans can readily draw boundaries that can be perceived on a map or a drawing, with one side of the boundary displaying one feature of flora, such as grasslands, and the other side forested lands (Table 19.3).

19.5 Transboundary Aquifers

In conducting a study of a very large aquifer system, the Rum-Saq Aquifer (Puri 1997), the author stated that, "aquifers know no boundaries" (except hydraulic ones) because it rapidly became clear that the area of study that initially was only several hundreds of square kilometres had to be extended to several thousand square kilometres—the hydrological boundaries that had been applied initially prevented the required level of understanding of the matter. However, the extended area that had to be included was affected by another type of boundary, the international boundary between two countries.

The experience gained from that study led the author to be a proponent of the study of transboundary aquifers. In the years since that first statement regarding aquifers knowing no boundaries, collaborative work with other disciplines, principally lawyers and humanists, has resulted in the UN having a adopted a resolution that encourages countries that possess transboundary aquifers to approach the issue by referring to Draft Articles prepared by the International Law Commission, grounded in legal formulations (Puri et al. 2007).

Laudable though these Draft Articles are in encouraging countries to cooperate (Puri 2003), what lessons can we draw when we come down to the level of an individual, say a farmer that uses ground water, which flows unseen and below ground across his boundaries?

A farmer and landowner works within a well defined parcel of land determined by ownership boundaries. Within this boundary area he is free to use the land as he pleases. In theory, the landowner has rights to all land within the boundaries to the centre of the earth….and if there is an aquifer under the land, then the farmer would not be wrong in considering this to be owned totally by him and his family.

The dilemma that we need to deal with is this: users of groundwater *do* know of their land boundaries but *not* of the boundaries of aquifers. Why does this dilemma matter, and how can it be resolved?

The discussion can has moved from the conceptualisation of boundaries, to their characterisation and their calculation. This is the scientific part. However, it does not deal with the dilemma fully, because we need to introduce the concept of common goods and values, and for these we again need to turn to those unseen and cultural boundaries and to the sense of sharing common values. If cultural norms give rise to boundaries which cause unacceptable harm to the whole community, say by overgrazing the common meadows, then we would transpose this cultural value also on the use of groundwater in a common aquifer. Then, as in the case of the meadow, we first need the community to be aware of and recognise their aquifer, which holds an invisible water resource.

This calls for management beyond the boundary, and it is a challenge as it calls for blurring the limits in which we work. We can no longer determine just the hydrogeology of an aquifer system, but also the hydro-economy of the community that uses it, we cannot then disregard the social norms of that community and we need to go beyond the catchment of the basin and of the mind. These are challenges that are being faced in some regions of the world. How effective they are for posterity, only time will tell.

Fig. 19.1 The distribution of national boundaries that overlie the Rum-Saq Aquifer system

19.5.1 Boundaries of Aquifers: Seen and Unseen

The Rum-Saq aquifer is only visible on the land surface in the southern part of the territory of Jordan and Saudi Arabia. The topography and the land features are shown in a satellite image in Fig. 19.1.

As illustrated in Fig. 19.2, the land territories of Jordan and western Saudi Arabia are underlain by 'unseen boundaries' that can be evoked by an analysis of the piezometric surface of the aquifers that occur in the region.

A hydrogeological analysis of the Rum Saq aquifer can be represented by a cross section that follows one of the flow lines of the piezometric surface—the flow line acts as a 'boundary' along which water can flow down gradient, but not across it. Figure 19.3 shows the dynamics of movement of water in the aquifer system. The cross section also represents the predicted response of water levels if pumping from the aquifer is carried out for 50 and 100 years.

For possible operations involving the pumping from aquifers, water levels would be drawn down and local changes in flow directions would take place. Boundaries associated with these flows would change in a dynamic manner. These changes are slow and take place gradually, thus the boundary evolution is gradual.

A synthesis of the information provided in Figs 19.1–19.3 can be summarised in a map as shown in Fig. 19.4. This two dimensional representation of the three dimensional conditions is demonstrated by the various boundaries that have been adopted to gain a good understanding of the transboundary aquifers. Firstly, geological boundaries indicate the area of the outcrop of the aquifers. Superimposed on

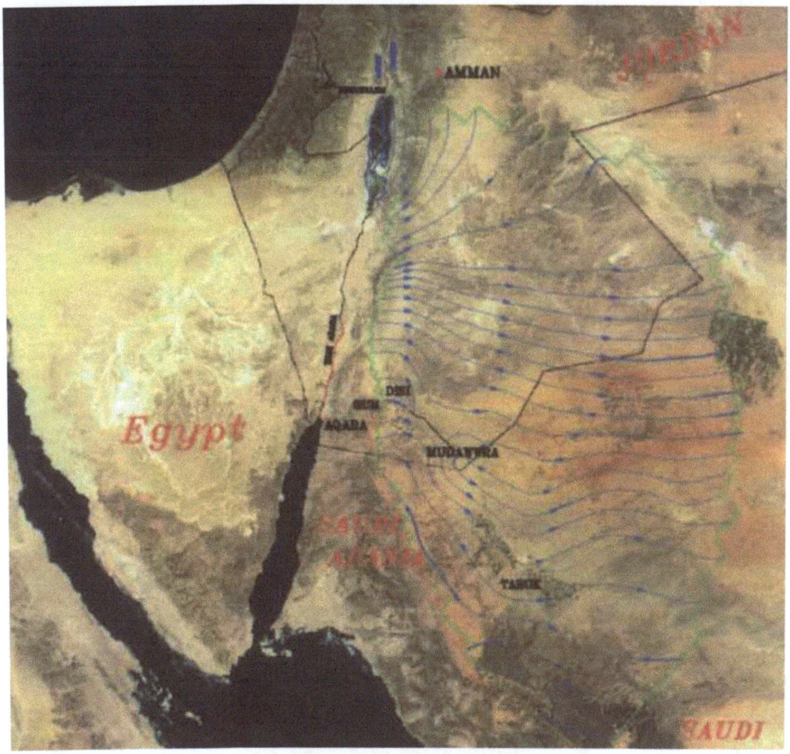

Fig. 19.2 The piezometric distribution of the aquifers underlying the territories of Jordan and western Saudi Arabia

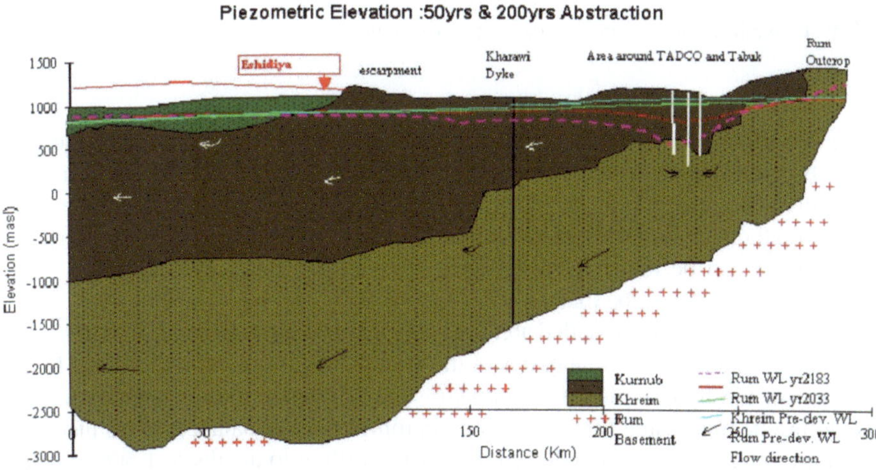

Fig. 19.3 Cross sectional representation of the aquifer system and some features reflecting change due to pumping over long periods

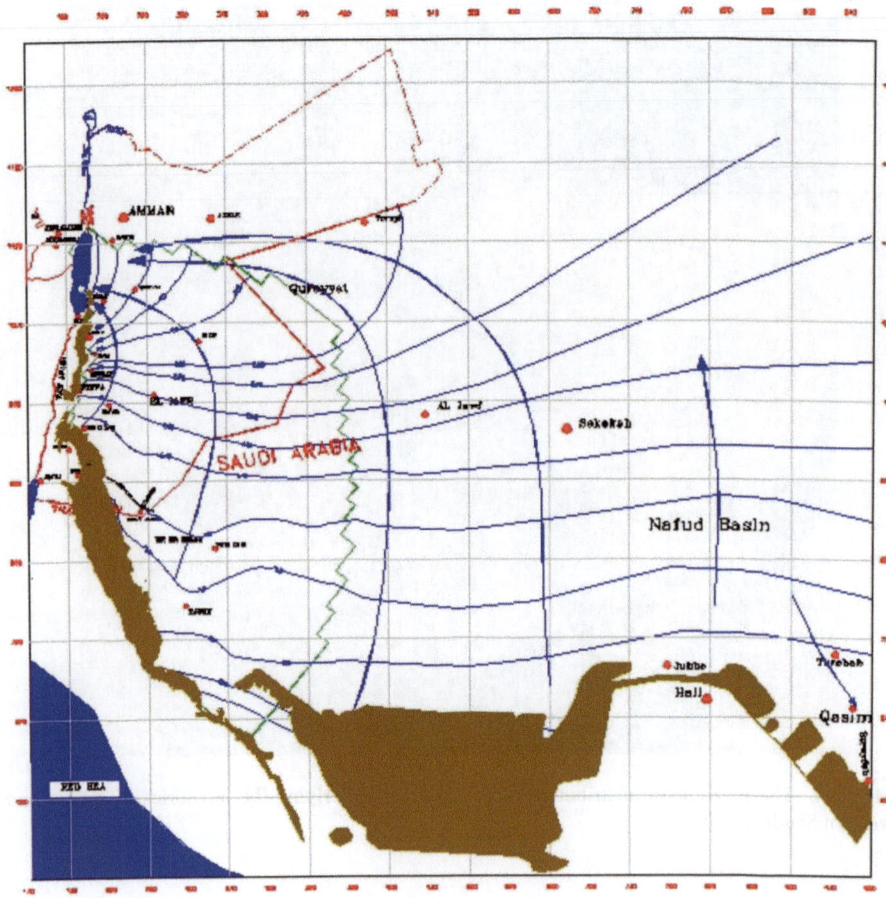

Fig. 19.4 Synthesis of the transboundary aquifer system

these are the piezometric lines and perpendicular to them can be found the flow lines. By following a selected flow line it has been possible to isolate a portion of the aquifer along which all of the water discharges to a deep inland sea, the Dead Sea. A mathematical model of the selected area can be constructed and a series of assessments can be made with this area—thus permitting a transboundary water resource assessment to be made.

The synthesis of the modelled area was only reached after several stages of assessment of the extent of the boundaries of the system, as illustrated in Fig. 19.5. In this assessment the hydrological limitations of taking a small area of the aquifer into modelling demonstrated that calculated appropriate water balances were incorrect. Field investigations were conducted, involving drilling to depths of 1,500 m below ground and these provided evidence that the model boundaries needed to be expanded until the phase 4 of the assessment gave the final conclusive understanding.

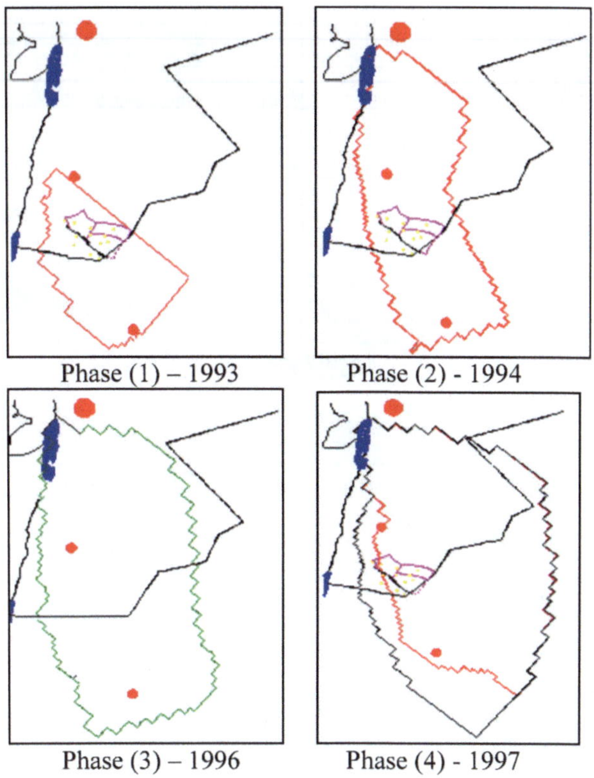

Fig. 19.5 Staged development of the modelled area in 4 phases (1993–1996)

19.6 Visualisation and Conceptualisations

The foregoing discussion demonstrates that almost all of the 'boundaries' that were discussed above are 'unseen'—as they occur in the conceptualisation of these systems. These boundaries cannot be physically observed, yet they are critical for the formulation of policy options that are needed for decisions on the way human societies operate. Yet, for transmitting the messages to those who need to grasp the issue, there is a need to provide visual and conceptual images. In the following two such visualisations are offered—one that visualises the social-political domain and the other that conveys the boundaries that are invoked within the aquifer systems.

While the hydrological boundaries and the appreciation of the physical system have been gained through the process described above, another aspect relates to the perception and understanding of the social and economic relations across national boundaries. Figure 19.5 captures a perception of the hydro hegemony in the region. The thesis that has been put forward by some suggests a three dimensional perception of the political economy—levels of cooperation and power balance are required. The 3-D plot suggests that the ideal position can be reached when there is high robustness in the political economy and sound technical assessments can be made.

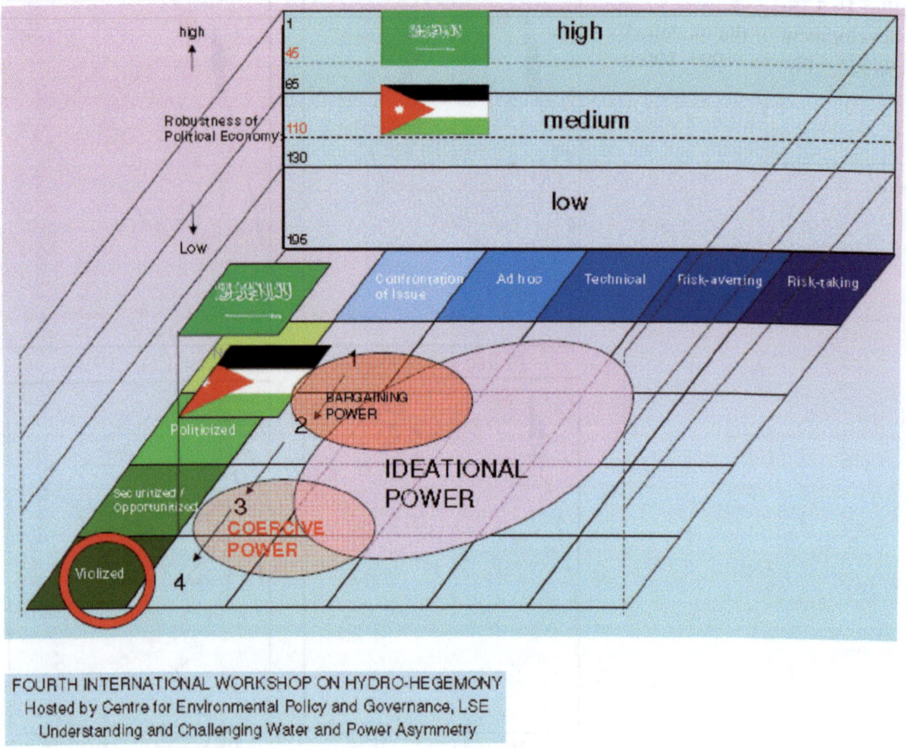

Fig. 19.6 Conceptualisation of the hydro-hegemony, political economy and power balance

While the above conceptualisation may be based on incomplete information, as it has been developed by external observers of the geopolitics of the region (The Centre for Environmental Policy and Governance, London School of Economics), the approach is novel and could have promise when applied in the context of full and reliable information on inter-country relations. Clearly there remains much to be done in gaining a better understanding of the cross-cultural relations that are separated by boundaries.

Visualisation of the dynamics of the aquifer system is also no easy task, since the bias of the specialist enters into the portrayal of the system. The 3-D block image in Fig. 19.6 would be clear to a trained hydrogeologist, but may not provide sufficient knowledge to the non specialist and possibly the policy maker. The 3-D cut away has been prepared from the physical perspective of a part of the territory of Jordan from point located high up in space, to the north east of the city of Amman. From this angle the Dead Sea is a prominent feature as it lies approximately 800 m below sea level and is therefore a natural drain to which all water in the region might flow—as ascertained by the direction of the flow of the river Jordan (not shown in the figure).

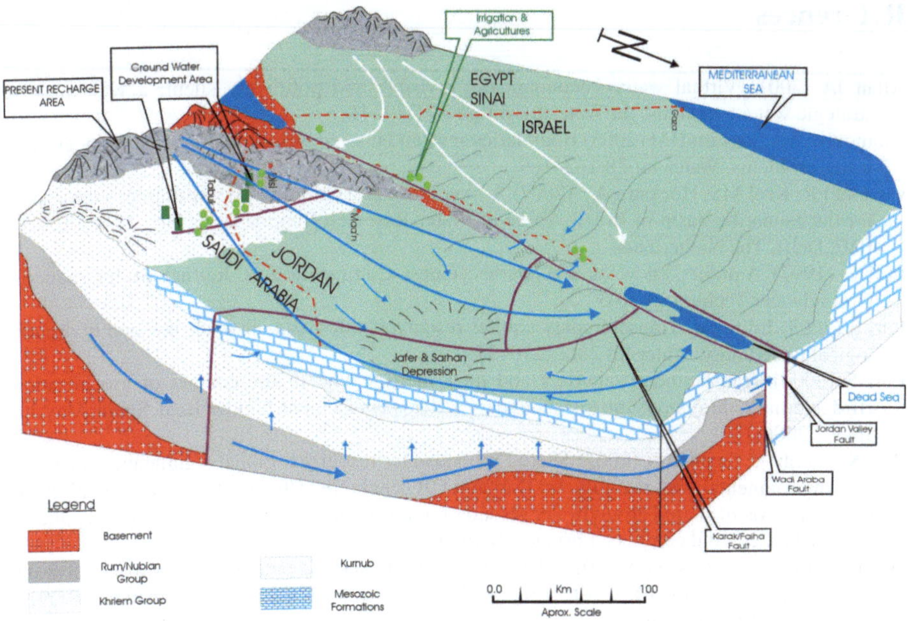

Fig. 19.7 The three dimensional physical assessment

The outcrop areas are the point at which groundwater, unseen and below ground, silently flows along flow lines that have been invoked based on field measurements of water levels in wells and then extrapolated to interpret the piezometric head distribution of the water (Fig. 19.7).

19.7 Concluding Remarks

This paper has taken a broad view of the 'boundaries' that are invoked in our daily lives to answer the question posed in the title of the Symposium 'The dilemma of boundaries—towards a new definition of the catchment'—which could imply that only the boundaries related to water pose a dilemma. In fact as discussed above there is a wider dilemma that applies across much of human endeavour—it would seem that we need to invoke 'boundaries' in practically all aspects of our daily lives. Some boundaries are seen and others are unseen. Clearly both types are significant—the seen boundaries occur in our local world; the unseen boundaries occur in the mind and in our perceptions. Perhaps the latter are much more important for us to address and fully appreciate since they are often the drivers of conflict.

References

Allan JA (2001) Virtual water—economically invisible and politically silent: a way to solve strategic water problems. Int Water Irrigation J 21:39–41

Blomquist WA, Ingram HM (2003) Boundaries seen and unseen: resolving transboundary groundwater problems. Water Int 28:162–169

van Oel PR, Krol MS, Hoekstra AY (2007) A river basin as a common-pool resource: a case study for the Jaguaribe basin in Brazil. Value of Water Research Report Series No. 24, UNESCO-IHE, Delft, The Netherlands

Puri S (1997) Aquifers know no boundaries, Guest Commentary in Journal of International Groundwater Technology, p 6

Puri S (2003) Editorial, Transboundary aquifer resources: international water law and hydrogeological uncertainty. Water Int 28(2):276–279

Puri S (2006) Transboundary aquifers and their management in the context of globalisation. IIIrd international symposium transboundary waters management, Ciudad Real, Spain, May 30–June 2 2006

Puri S, Aureli A, Stephan R (2007) Shared groundwater resources: global significance for social and environmental sustainability. In: Darnault CJG et al (eds) Chapter 1, Overexploitation and contamination of shared groundwater resources: management, (bio)technological, and political approaches to avoid conflicts. Springer, Netherlands

Warner J, Lulofs K, Bressers H (2010) The fine art of boundary spanning: making space for water in the East Netherlands. Water Altern 3(1):137–153, http://www.nature.com/climate/2009/0910/full/climate.2009.92.html

Zeitoun M, Warner J (2006) Hydro-hegemony—a framework for analysis of transboundary water conflicts. Water Policy 8:435–460

URL

Nature Reports Climate Change. Published online: 23 September 2009, doi:10.1038/climate.2009.92, http://www.nature.com/climate/2009/0910/full/climate.2009.92.html

Chapter 20
Sustainable Management of Groundwater Resources for 700,000-Plus Residents: A Practical Example of the Transboundary Management of Groundwater Resources in the Kumamoto Area, Japan

Jun Shimada

Abstract The city of Kumamoto and its surrounding areas have a population of more than 0.7 million, and unique in that the prefecture-governed city boasts a water supply that is exclusively supported by an abundant groundwater system. The recent expansion of urban land-use areas in and around the city has resulted in a decrease in the volume of regional groundwater resources. This is despite the efforts of local governments to maintain stable groundwater recharge rates and the economical use of water. In addition, groundwater contamination, caused primarily by agriculturally derived nitrate, is increasing. Public education is necessary to maintain environmentally stable regional groundwater resources in terms of both quantity and quality. This paper describes the ongoing program conducted by the city of Kumamoto for this purpose.

Keywords Abandoned paddy fields • Artificial recharge • Land use change • Local city government • Transboundary groundwater management • Volcanic aquifer

20.1 Introduction

Groundwater is an indispensable resource for the city of Kumamoto, the third largest city on the island of Kyushu, Japan. The city is the prefectural capital, and groundwater is provided to its 700,000-plus residents at each water tap. A natural spring fills Lake Ezu, in Suizenji Park, and many other locations within the city. There are few other cities of this size in Japan that rely entirely on groundwater (Fig. 20.1).

J. Shimada (✉)
Graduate School of Science and Technology, Kumamoto University,
2-39-1 Kurokami, Kumamoto 860-8555, Japan
e-mail: jshimada@sci.kumamoto-u.ac.jp

M. Taniguchi and T. Shiraiwa (eds.), *The Dilemma of Boundaries: Toward a New Concept of Catchment*, Global Environmental Studies,
DOI 10.1007/978-4-431-54035-9_20,

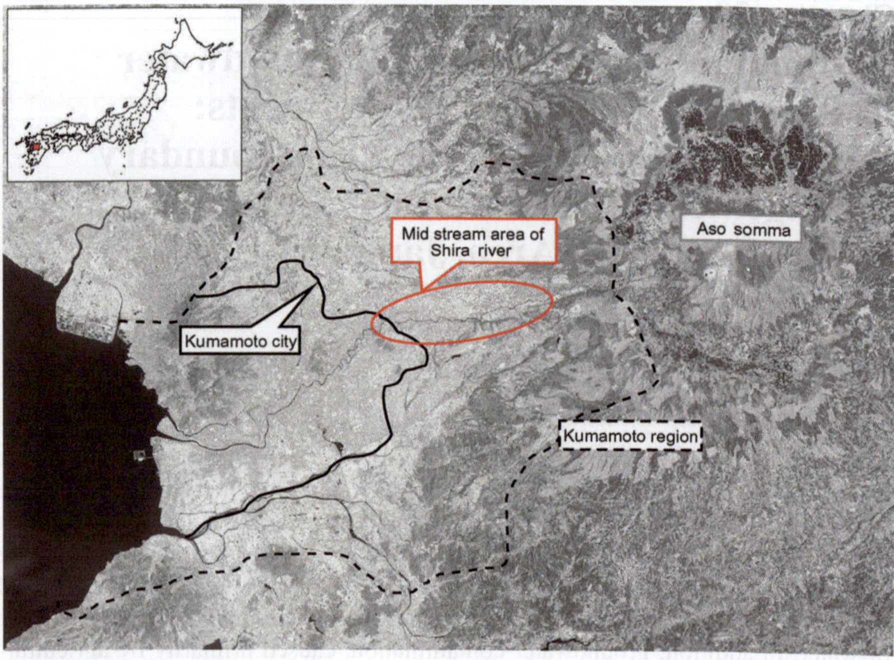

Fig. 20.1 Location of the study area

Kumamoto has thrived thanks to its precious and abundant groundwater, and the task for the citizens of Kumamoto is to maintain this rich and clean groundwater for future generations. However, groundwater levels have been decreasing recently, as has the discharge volume from Lake Ezu, with a decline evident in groundwater recharge. Before the supply is depleted, Kumamoto citizens and corporations that utilize the groundwater must acknowledge its value and take measures to preserve their groundwater resources.

20.2 Basic Aquifer Structure of the Study Area

Figure 20.2 shows the surface geology of the area. Most of the study area is covered by pyroclastic deposits that were created by the four major eruptions of the Mt. Aso caldera that occurred between 0.26 Mya and 0.09 Mya. There are two aquifer systems: the unconfined aquifer (No. 1 aquifer) and the confined aquifer (No. 2 aquifer). The No. 2 aquifer is the major groundwater resource in the Kumamoto region, as shown in Fig. 20.1. The huge groundwater aquifer basin in this region was created by relatively high local precipitation (2,200 mm/year) and the permeability of the pyroclastic deposits.

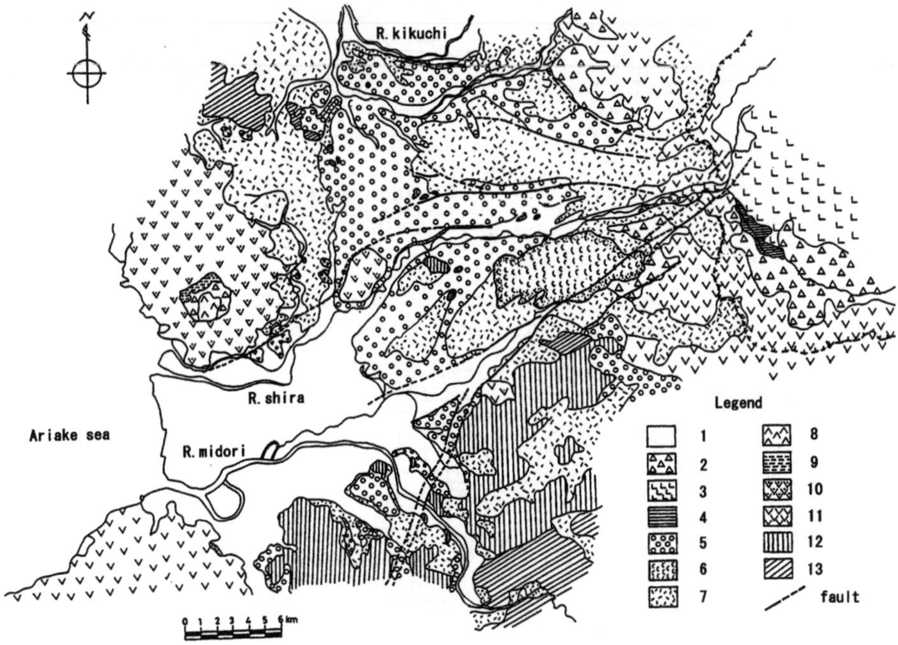

Fig. 20.2 Surface geologic map of the study area. 1: Alluvium, 2: Cliff deposits, 3: Aso crater hills, 4: Kufuno deposits, 5: Terraces, 6: Takayubaru lava, 7: Pyroclastic flow deposits, 8: Ichinodake lava, 9: Yoshino deposits, 10: Pre-Aso volcanic rocks (Kinpo volcanic rocks), 11: Pre-Aso volcanic rocks (others), 12: Cretaceous sedimentary rocks, and 13: Metamorphic rocks

Figure 20.3 shows the potential groundwater distribution of the No. 2 aquifer in October 1993, which is the annual high water season. In the area shown by the dashed oval, annual groundwater levels fluctuate by more than 10 m, indicating that the majority of recharge occurs in June when there is heavy rainfall during the summer rainy season. The lack of a lacustrine deposit layer separating the unconfined and confined aquifers enables rainwater and irrigation water to directly recharge the No. 2 aquifer system. This hydro-geologically unique area is termed the groundwater pool for this aquifer, and is considered its most important recharge area. The effect of this recharge area within the groundwater flow system has been clearly shown using environmental stable water isotopes (Kosaka et al. 2002).

The groundwater recharged in this area flows southwest along the large arrows in Fig. 20.3, with springs filling Lake Ezu in Suizenji Park, and many other spring locations within the city. The travel time of this groundwater flow is 10–20 years (Kagabu et al. 2010).

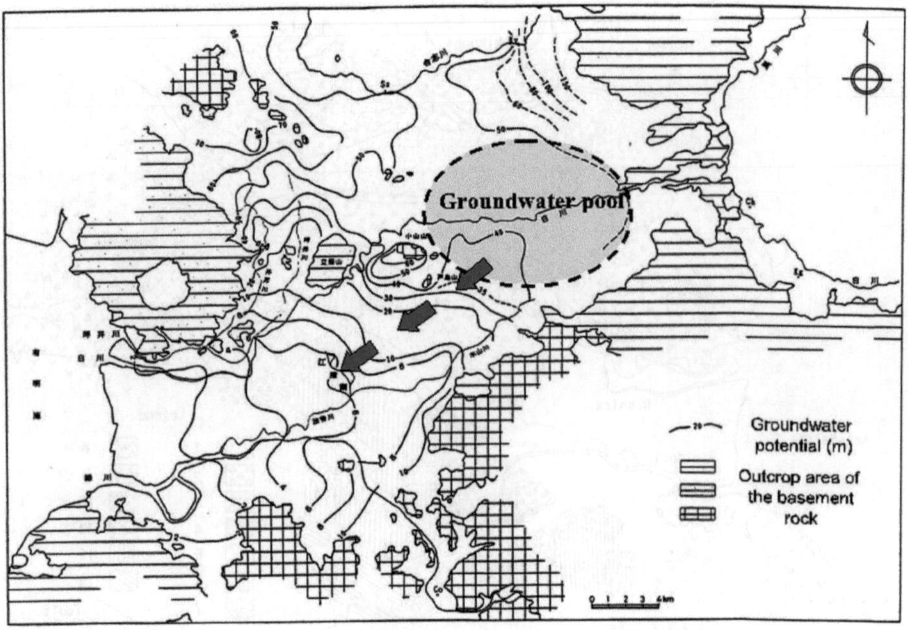

Fig. 20.3 Groundwater potential distribution of the No. 2 aquifer in the Kumamoto area (Oct. 1993)

20.3 Volume of Groundwater in the Area

The city began to measure groundwater levels in 1986 via a wells observation network. Since that time, the levels at both discharge and recharge areas have clearly decreased. A marked decline in the water level is evident at the Ozu observation well, situated in the mid-stream area of the Shira River (Fig. 20.4). The discharge rate of spring water in Lake Ezu, Suizenji, has diminished by approximately 20% over the last decade, from approximately 500,000 m^3 per day in 1990 to 400,000 m^3 per day in 2001 (Fig. 20.5). In the 1950s, the discharge rate was approximately 1,000,000 m^3 per day.

The "balance" of groundwater, calculated by deducting the outflow from groundwater recharge, is shown as a deficit, and the "deficit" accumulates annually. To improve this imbalance, the volume of groundwater withdrawal must be reduced and the recharge of groundwater must be increased.

The decline in the volume of groundwater withdrawn in Kumamoto city, which is shown in Fig. 20.6, is mainly due to a significant decrease in the volume of groundwater extracted for industrial and agricultural purposes. These decreases are the result of reduced extraction efforts by local industries and by the agricultural diversion of paddy fields to alternative crops; however, there has been a slight increase in city water use, which now accounts for more than 70% of total groundwater consumption.

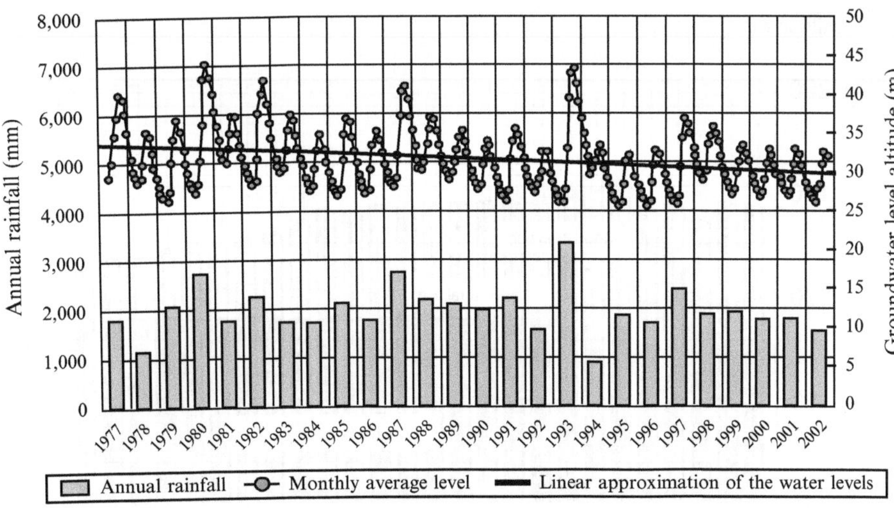

Fig. 20.4 Annual changes in groundwater levels at Ozu in the mid-stream area of the Shira River. *Source*: From data provided by the Ministry of Land, Infrastructure and Transport

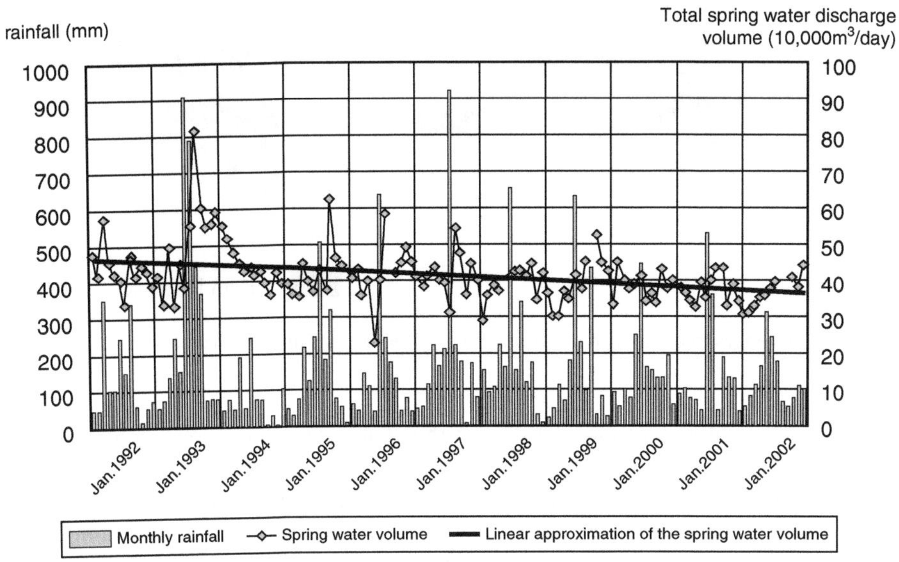

Fig. 20.5 Annual changes in spring discharge at Lake Ezu. *Source*: From surveys conducted by Kyushu Tokai University

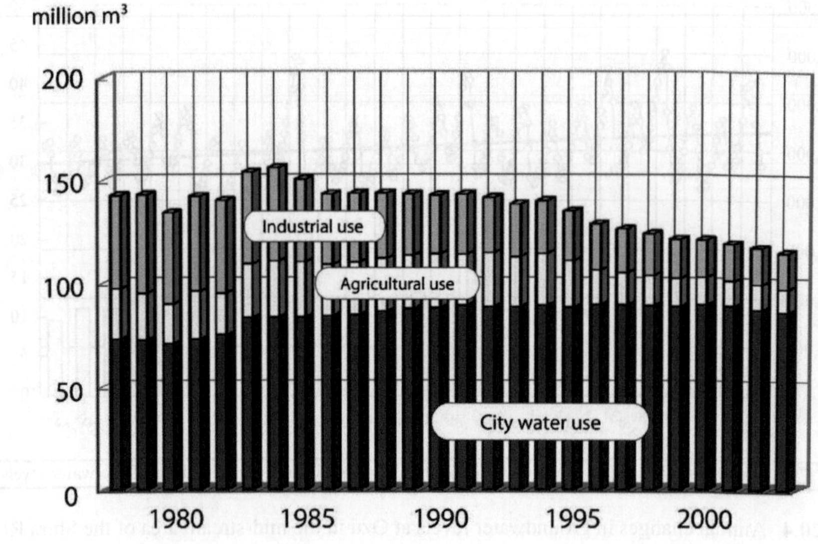

Fig. 20.6 Annual changes in groundwater withdrawal uses in Kumamoto city

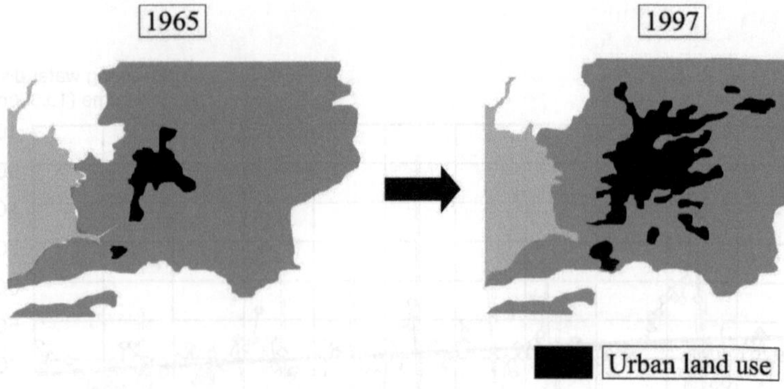

Fig. 20.7 Changes in urban land use from 1965 to 1997 in the Kumamoto area

Figure 20.7 illustrates the changes in urban land use in the Kumamoto area from 1965 to 1997. The growth of the urban population has resulted in farmland being redeveloped for urban use, thus reducing the area available for infiltration. This recent change in land use is the principle reason for the decreasing groundwater recharge rate in the study area. Urban land use has expanded beyond the city to outlying areas that previously functioned as the recharge area for aquifer No. 2.

Of particular importance is the marked decrease in the number of rice paddy fields situated in the mid-stream area of the Shira River. Figure 20.8 shows the changes in non-agricultural land use in this area from 1930 to 2002. Non-agricultural land has been gradually expanding over the last 70 years. In addition, the land use

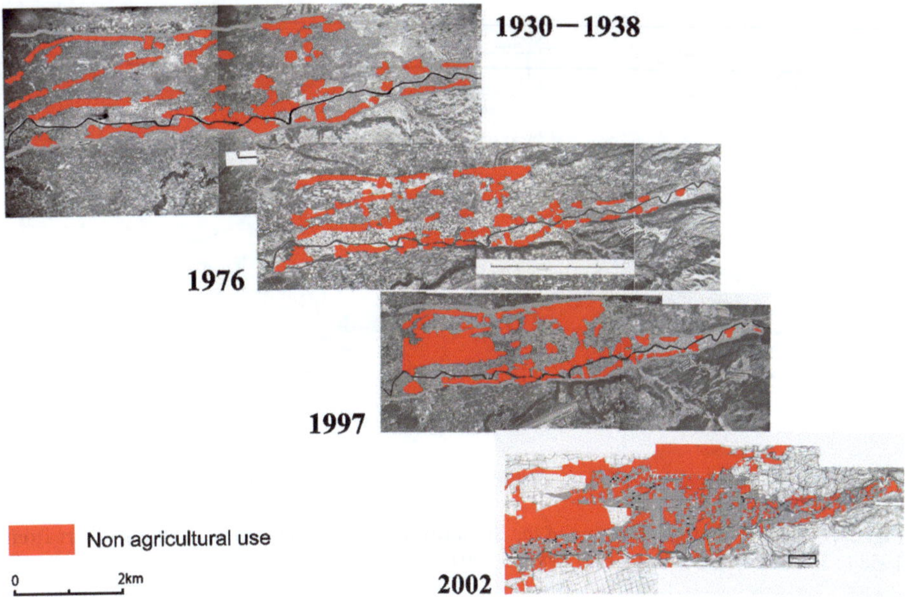

Fig. 20.8 Changes in non-agricultural land use in the mid-stream area of Shira River from 1930 to present (Ichikawa 2004)

pattern of farmland has changed from rice paddy fields to dry vegetable fields, in response to national policies directed at the control of rice production and to reduce the need for irrigation during the rice-growing season.

As mentioned previously, the mid-stream area of the Shira River is an effective recharge area for the No. 2 aquifer. When local farmers used this land for cultivating rice, considerable volumes of irrigation water were required because of water leakage from the paddy fields. The average daily leakage from the rice fields in this area ranged between 100 and 200 mm, which is approximately ten times greater than that associated with normal Japanese paddy rice fields, and is caused by a porous layer of coarse gravel existing in this area. While this high leakage rate was detrimental for rice farmers, it was beneficial for groundwater recharge in the region.

As shown in Fig. 20.8, approximately half of the mid-stream area of the Shira River once used for rice paddy fields has been converted into residential areas or dry vegetable farmland; therefore reducing the volume of water recharging the No. 2 groundwater aquifer. The main reason for this change in land use is Japanese agricultural policy, which limits rice production to maintain higher prices. Interestingly, this reduction in rice cultivation has had an effect on regional groundwater resources.

Further evidence of a reduction in paddy irrigation is illustrated by the long-term river discharge trends of the Shira River. Figure 20.9 shows an increase in the discharge of the Shira River during the last 40 years, obtained by comparing two

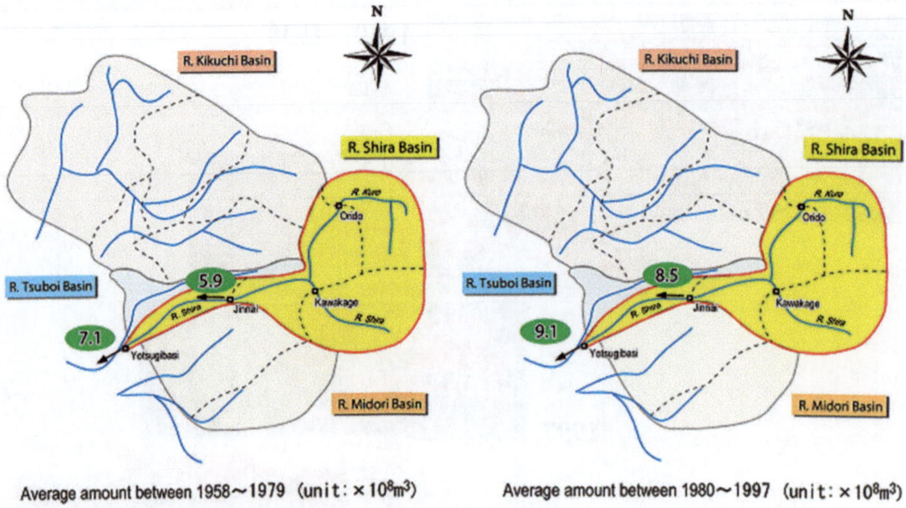

Fig. 20.9 Change in river discharge of the Shira River during the last 40 years: the 20 years from 1958 to 1979 and the 20 years from 1980 to 1997

representative periods: the 20 years from 1958 to 1979 and the 20 years from 1980 to 1997. The precipitation records for the Kumamoto area do not show any significant increases during the last 100 years. Unlike the Shira River, the three other major river systems in the area showed no significant increases in discharge volumes for the same period (Shimada et al. 2009). The increasing discharge volumes of the Shira River are believed to be the result of the abrupt reduction of paddy irrigation, thereby reducing the intake of river water via irrigation channels, ultimately returning the river to its natural flow rate.

20.4 Integrated Transboundary Groundwater Management

To preserve Kumamoto's abundant groundwater resources for future generations, recharge of the groundwater reservoir must be increased, alongside a reduction in groundwater withdrawals. To achieve these objectives, it will be necessary for the Kumamoto prefectural and city governments to implement specific measures.

The most effective means of increasing the groundwater reservoir is through collaboration with local farmers. To recharge their groundwater reservoirs, farmers can implement measures to increase water recharge through rice paddies in the midstream areas of the Shira River, where the largest volume of recharge occurs. In addition, an annual conference will be held to discuss development initiatives and the preservation of water reserves. The "Conference for Utilizing Rice Paddies for Groundwater Recharge along the mid-Stream Shira River" will be attended by local communities from the land improvement districts, the Japan Agricultural

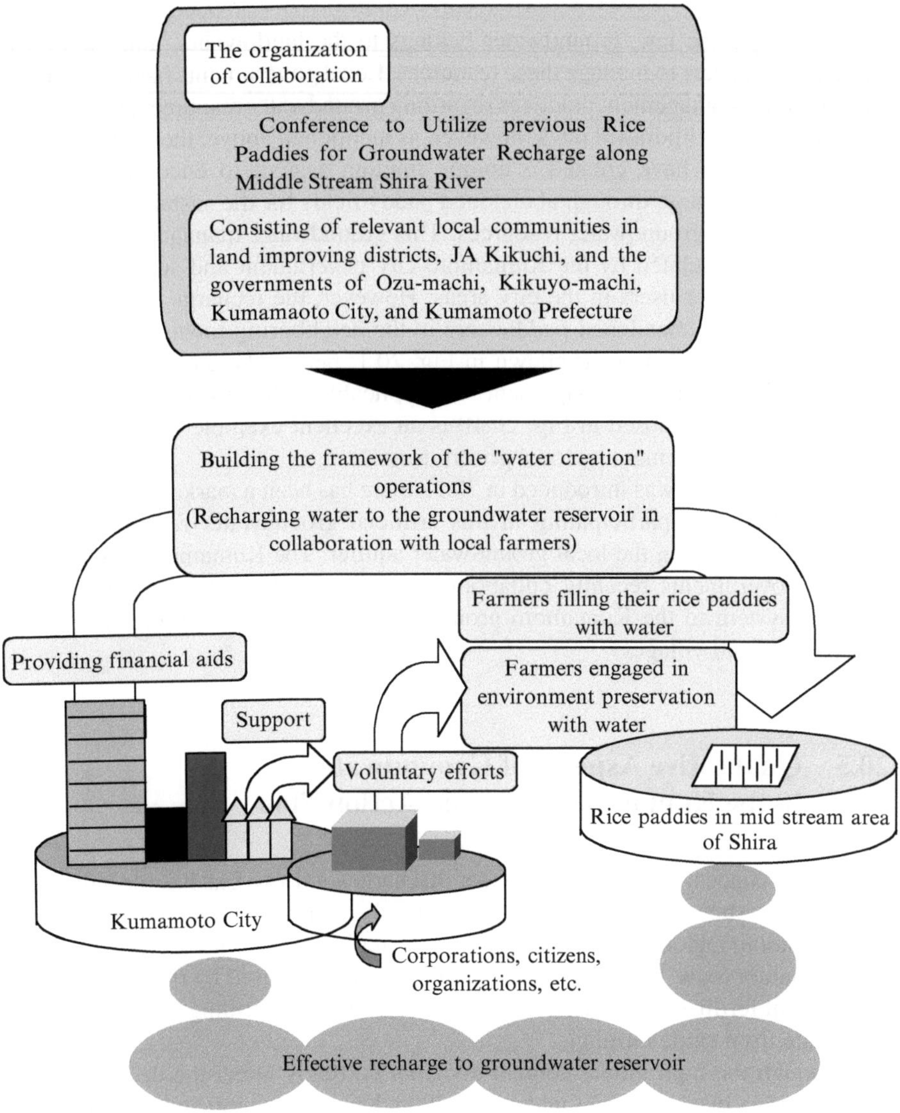

Fig. 20.10 Schematic view of the creation of groundwater resources through rice paddies

Cooperatives (JA), and the governments of the towns of Ozu and Kikuyo, Kumamoto city, and Kumamoto prefecture. Local organizations will collaborate to discuss plans to fill the abandoned rice paddies with river water. Because the local rice paddy farmers have certain surface water rights, they will still be allowed to utilize river water legally for irrigation of their paddy lands, and even for non-rice production. Financial assistance will be provided to local farmers who voluntarily fill their abandoned paddies with river water (Fig. 20.10). Voluntary measures to promote the "creation of water" by corporations and citizen groups will also be supported.

Under Japanese law, groundwater belongs to the landowners, and there is no unified national law to manage these resources. Local governments have only implemented minor management practices regarding groundwater resources, for example, the recording of withdrawal rates; however, as mentioned above, the prefectural and city governments have created a unique funding system to encourage artificial groundwater recharge through abandoned paddy fields for the sustainable management of regional groundwater resources. This groundwater management system is proposed and regulated by the Kumamoto city government and supported by the major groundwater users in the city areas. However, the recharge areas that most effectively utilize abandoned paddies are in the neighboring towns, outside of the Kumamoto city boundary, as shown in Fig. 20.1; hence, the term 'transboundary groundwater resources management' is applicable. The artificial groundwater recharge system proposed in Fig. 20.10 is an excellent example of transboundary groundwater management by local government.

Since the system was introduced in 2004, there has been a marked increase in the number of farmers participating in this artificial groundwater recharge, thereby effectively recharging the local groundwater aquifer. The Kumamoto city and prefectural governments recently collaborated to expand this artificial groundwater recharge system to the Kumamoto groundwater basin scale, comprising 11 local cities, towns and villages.

20.5 Qualitative Aspects of Groundwater in the Kumamoto Area and Action Plan

The nitrate concentration in spring water discharged to Lake Ezu has recently risen. Figure 20.11 shows the distribution of nitrate concentrations in the groundwater of the Kumamoto region. The northern area of the region exhibits a high level of nitrate contamination (over 10 mg/l NO_3^- contents), generally caused by the excessive use of chemical fertilizers in intensive hot-house watermelon farming and through animal waste from cattle farming.

Although the high nitrate content does not currently affect the deep drinking-water aquifers in a significant manner, in 2005, Kumamoto prefecture established a nitrate reduction plan to monitor the nitrate concentration in groundwater over a 10-year period.

The action plan includes:

(a) Reduction of spray time and total volume of chemical fertilizers.
(b) Replacement of chemical fertilizers with organic fertilizers ("ECO farming").
(c) Construction of purification plants for animal waste, depending on the size of the cattle farm.
(d) Prohibition of pit disposal methods for animal waste.
(e) Effective utilization of animal waste as organic fertilizer.
(f) Development of a public sewerage system.
(g) Improving the environmental ethics of local farmers.

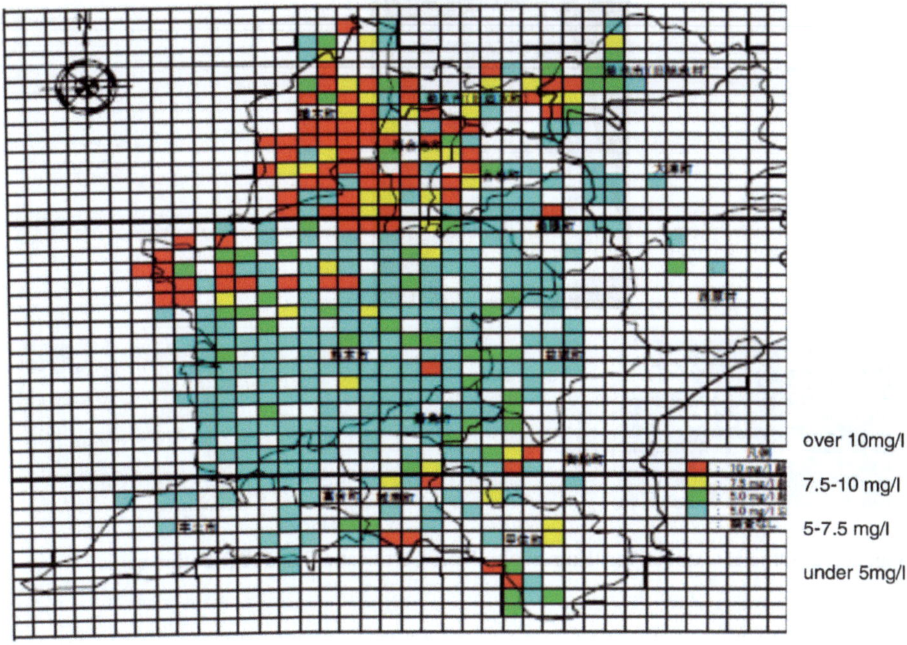

Fig. 20.11 Nitrate concentration in groundwater in the Kumamoto area (2000)

20.6 Conclusions

The Kumamoto city government has recently established a unique transboundary groundwater management system using abandoned rice paddies. It has proven effective in maintaining the purity of local groundwater for the area's 700,000-plus residents and for future generations. It is imperative that this continues because every household in the city depends on groundwater alone, a concern that local government has taken seriously. High levels of nitrate found recently in groundwater are a further issue of concern that has resulted in the establishment of a new action plan by the local prefectural government to reduce nitrate levels in groundwater.

To effectively promote these groundwater management plans, the understanding and cooperation of the residents is paramount. In 2008, the city government was awarded the Japan Water Prize for their long-term efforts to maintain local groundwater resources. Upon receiving this award, the city government established 'The Official Water Certification System' (Fig. 20.12), designed to assist in educating residents about groundwater and to help them understand the importance of protecting these water resources. It is hoped that this system will encourage citizens to become involved in the effective management of their local groundwater resources.

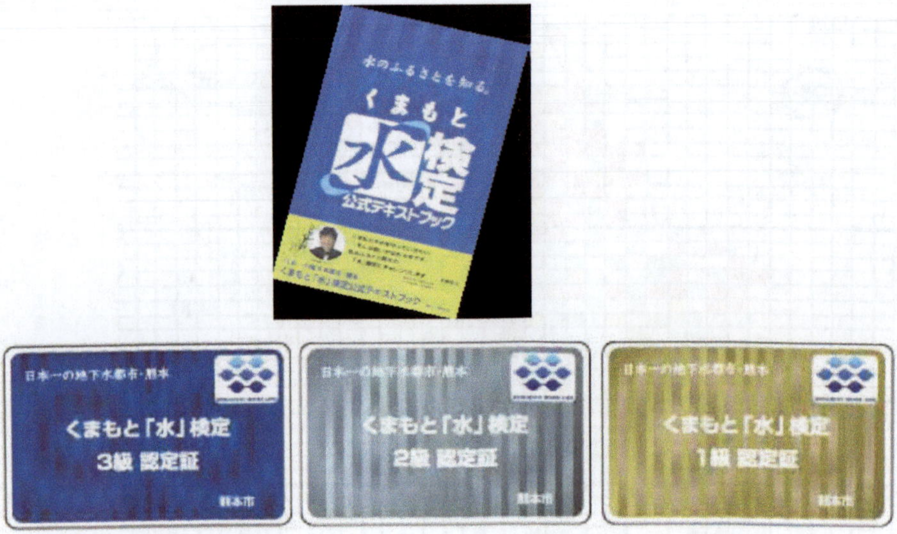

Fig. 20.12 Official textbook for 'Kumamoto Water Certification' and registration cards

References

Ichikawa T (2004) The recent drawdown of groundwater discharge at Kumamoto area and the importance of the mid-stream area of River Shira. Proceedings of symposium on healthy groundwater cycle from the aspect of quality and quantity sponsored by the Japanese association of groundwater hydrology, 21–28

Kagabu M, Yamaguchi, Shimada J (2010) Residence time of groundwater of the Kumamoto area by using repeated Tririum measurement. Proceedings of annual spring meeting of Japanese Association of Groundwater Hydrology. No. 04, Tokyo, May 2010

Kosaka H, Shimada J, Sakamoto M (2002) The origin of the groundwater in the Kumamoto area by using a stable isotope ratio. Proceedings of annual spring meeting of Japanese Association of Groundwater Hydrology, 140–143

Kumamoto City (2004) Kumamoto city's plan for preserving its groundwater reservoir, p 37

Kumamoto Prefecture & Kumamoto City (1995) Integrated groundwater survey report of the Kumamoto area, p 122

Shimada J, Okamura N, Kagabu M (2009) The cause of groundwater decline shown in the long-term river discharge record. Proceedings of joint international convention of 8th IAHS Scientific Assembly and 37th IAH congress, JW1–2/6, Hyderabad, 6–12 September 2009

Part VI
Conclusion

Chapter 21
The Dilemma of Boundaries in Environmental Science and Policy: Moving Beyond the Traditional Watershed Concept

Takeo Onishi, Makoto Taniguchi, Takayuki Shiraiwa, Takahiro Endo, and Yasunori Hanamatsu

Abstract We are repeatedly drawing and redrawing various boundaries around us. The function of these boundaries is not only to divide people but also to bound people. Thus, boundaries are essential for daily life. The articles collected in this book shed light on human made boundaries from the viewpoint of water related problems. Amongst various water related environmental problems, the concept of the watershed has been the central dogma in water management schemes, and it also presents boundary dilemmas. Critically considering the conventional watershed concept, we tried to identify the limitation inherent in the watershed concept. As a result we found that in some cases conventional watershed boundaries work well, but in some cases, they may not work well at all. Since global and local water are inevitably inter-connected and the structure of inter-connectedness is changing, boundaries should change according to the structural change of inter-connectedness. Thus, the ability of our social systems to adapt to newly emerging boundaries is important. The same kind of idea can be found in the field of ecology and information

T. Onishi (✉)
Faculty of Applied Biological Sciences, Gifu University, 1-1 Yanagido, Gifu 501-1193, Japan
e-mail: takeon@gifu-u.ac.jp

M. Taniguchi
Research Institute for Humanity and Nature, 457-4 Motoyama, Kamigamo, Kita-ku, Kyoto 603-8047, Japan

T. Shiraiwa
Institute of Low Temperature Science, Hokkaido University, Kita-19, Nishi-8, Kita-ku, Sapporo 060-0819, Japan

T. Endo
Graduate School of Life and Environmental Sciences, University of Tsukuba, 1-1-1 Tennoudai, Tsukuba, Ibaraki 305-8572, Japan

Y. Hanamatsu
Slavic Research Center, Hokkaido University, Kita-9, Nishi-7, Kita-ku, Sapporo 060-0809, Japan
e-mail: hanamatsu@slav.hokudai.ac.jp

M. Taniguchi and T. Shiraiwa (eds.), *The Dilemma of Boundaries: Toward a New Concept of Catchment*, Global Environmental Studies, DOI 10.1007/978-4-431-54035-9_21, © Springer Japan 2012

technology. The common theme found in both fields is 'management'. Though 'management' has not been thought of as a true scientific objective, particularly in the natural sciences, these sciences should, with the aid of the social sciences, include 'management' as a challenging objective.

Keywords Boundaries • Dilemma • Ecology • Information technology • Management • Watershed

21.1 Dilemma of Boundaries

Readers can find that the articles collected in this book shed light on human made boundaries from the viewpoint of water related problems. Firstly, water-related environmental problems induced by human made boundaries are summarized in Sect. 21.2. Then, in the Sect. 21.3, we critically consider the traditional watershed concept. Amongst the various boundaries invoked in environmental management, the watershed boundary seems to reflect the natural continuity of water, and thus it appears to have a rigid scientific basis. Actually, the concept of the watershed has been the central dogma in water management schemes, and it also presents boundary dilemmas. Critically considering the conventional watershed concept, we tried to identify the limitation inherent in the watershed concept. Lastly, in the Sect. 21.4, we discussed how we can step forward toward a new watershed concept that will enable better governance of water-related problems.

21.2 Water-Related Environmental Problems Induced by Human Made Boundaries

When the continuity of water is interrupted by human-made boundaries, water related environmental problems can occur. Problems relevant to water induced by human-made boundaries can generally be classified into two categories: problems related to physical boundaries (Class 1) and problems related to functional/meaning boundaries (Class 2). In other words, we can find Class 1 boundaries in physical space, but can't find Class 2 boundaries in physical space. Examples of Class 1 boundaries addressed in the book include the division of river continuity by dams (Matsuda), the interruption of river/ocean continuity by weirs (Matsuda), and the overexploitation of trans-boundary groundwater (Jarvis and Puri). Examples of Class 2 boundaries referred to in the book were upper–lower reach conflicts involving river waters and materials (Endo), fragmentation of laws between surface water and groundwater (Endo), higher–lower government conflicts (Backer), and sectionalism (Matsuda). Though Class 1 and Class 2 boundary issues are closely interweaved and interdependent, most boundaries categorized into Class 2 boundaries are deeply related to structures of human societies.

21.3 Dilemma of Watershed

Since Hardin (1968) pointed out the tragedy of the commons, the commons has been and remains one of the central issues in environmental management. Presently, the central issues concerning commons include how to effectively manage commons, what kinds of institutions allow for the sustainable use of commons, and how to generate new commons in cases where old systems have decayed or been replaced.

The watershed is a very effective framework for promoting the establishment of commons to overcome water-related environmental problems. Actually, the notions of Integrated Watershed Management (IWM) and Integrated Water and Land Management (IWLM) stem from deep reflection on the problems induced by human made boundaries. Since the 1990s, IWM and IWLM have become standard approaches for water governance around the world, though there remain many challenges to be tackled. Here, the definition of watershed is a conventional one: the area that includes both streams and rivers that transport water as well as shape the land surfaces from which water drains.

Though we tend to consider a watershed as if it is a closed system, no watershed is closed at all. Thus, the chapters in this book focus especially on the open system aspect of watersheds. Parts I and II discussed watersheds from the viewpoint of the natural sciences. In Chapter 4, Taniguchi pointed out that urbanization and global warming enhance heat deposition below ground to depths of over 100 m. In Chapters 8 and 9, arguments were raised by Shiraiwa and by Onishi concerning nutrient flow. When we focus on nutrients such as nitrate or phosphorous, the area of the watershed expands to include the coastal zone. In the case of nutrients such as iron, the watershed area can be extended further to include the open ocean. These studies revealed the influence of land on ocean. On the other hand, in Chapter 10, Murota introduced a very interesting case involving ocean to land nutrient flow mediated by living organisms such as anadromous fish. Figure 21.1 shows the classification of these examples by the spatio-temporal scale of each phenomenon. In summary, when we focus on the flows of materials, energy, living organisms and human activities, we see that they extend beyond the range of the traditional watershed.

Moreover, several authors in the book pointed out the importance of boundaries related to socio-economic considerations in Parts IV and V. In Chapters 14 and 19, Jarvis and Puri pointed out the importance of economic linkages, and in Chapter 16, Ieda pointed out the importance of cultural homogeneity. Through the collected articles in the book, we can conclude that though the watershed is a useful unit for the governance of water-related problems, we should recognize that watersheds are at the same time temporal boundaries. Indeed, the concept of land–ocean interactions in coastal zones (Leeks and Jarvie 1998) is a good example of the growing awareness of the connectedness between watersheds and coastal areas. Thus, we can call this situation the 'dilemma of watersheds'.

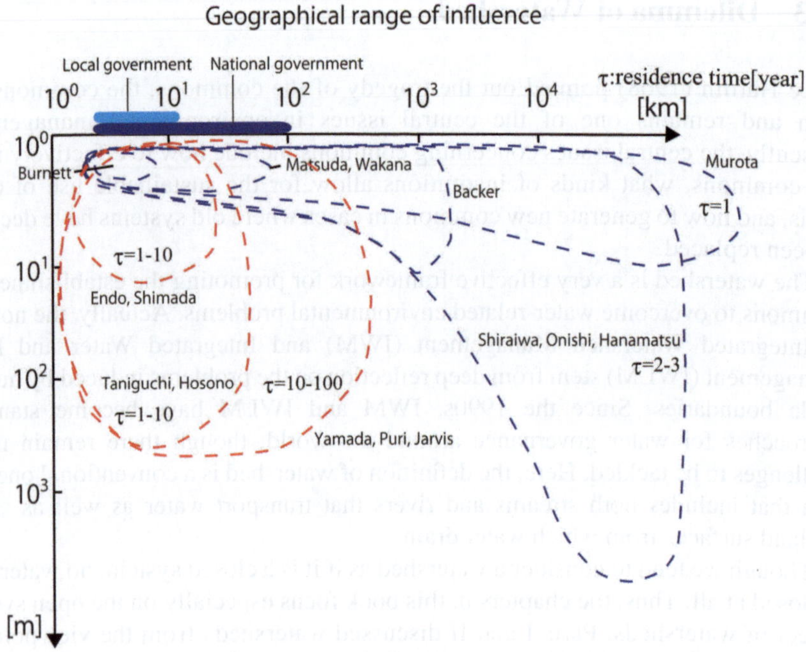

Fig. 21.1 Geographical scale and inherent time scale of various water related processes

21.4 How Should We Cope with the Dilemma of Watersheds?

21.4.1 Importance of the Inter-connectedness of Global and Local Water

In the last section, we reached the temporal conclusion that multi-layered something (heat/energy/living organisms) -sheds simultaneously overlie each watershed. Of course, we can find much evidence supporting the notion that local water circulation in different areas itself is connected at the global scale. The concept of virtual water (Allan 1999; Oki and Kanae 2008) insists that any water circulation in a local watershed is more or less influenced by activities in other parts of the world through agricultural commodities trading between countries. In addition, the extent to which countries influence each other's natural water circulation has now been quantitatively evaluated (Dirmeyer et al. 2009; van der Ent et al. 2010). Moreover, RIHN (2006) categorized the various forms of value assigned to water resources according to spatial scale (Fig. 21.2). The figure is the modified version of a figure originally depicted by Turton et al. (2005). In the first stage of water resource development in the last century, water was seen as a material resource but not as an economic resource. Only recently has economic value been added to water resources for the sake of management. There is also a recent movement to consider the social aspects

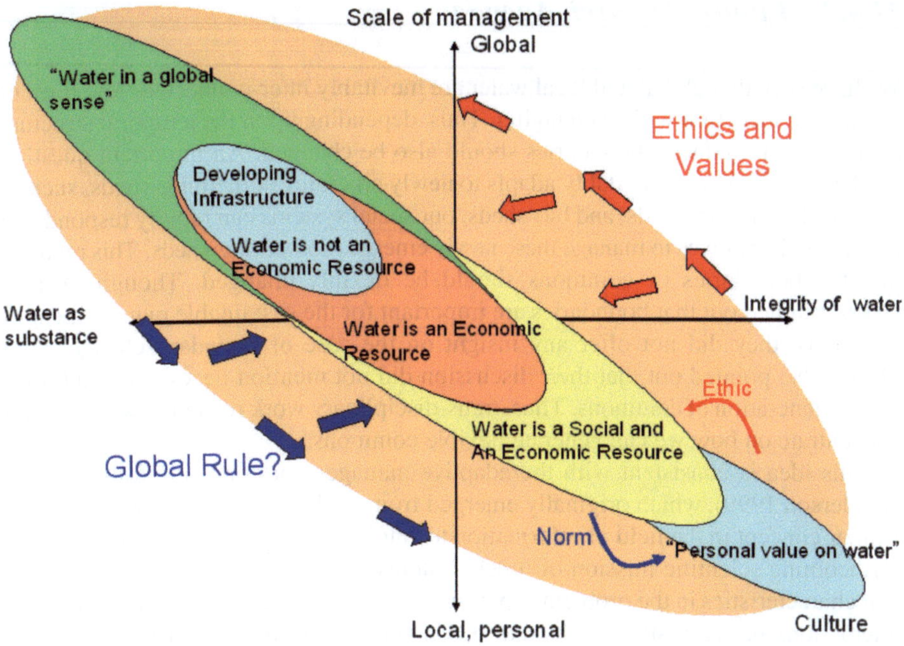

Fig. 21.2 Schematic diagram of various water values depending on the spatial scale (cited from RIHN 2006)

of water. The figure includes the global aspect of water as H_2O circulation and the personal aspects of water, which are very culturally- and individually-specific. Interestingly, the global aspect of water is dependent on individual perspectives, so they are inter-connected by ethics and norms. It seems that wiser use of water cannot be accomplished without such a global point of view.

Rockström et al. (2009) proposed the concept of planetary boundaries. They define boundaries as the safe limits within which the earth system can remain stable. In addition, they quantitatively estimated definite values in nine important domains that are critical for the earth system. In other words, human activities should not result in impacts that in aggregate violate these planetary boundaries; otherwise, the earth system will be very unstable. Environmental regulations in each country should be developed in the context of global environmental stability in the near future. This has already started in the fields of global warming and biodiversity.

Thus, it seems likely that incorporating the interconnectedness of human activities at multiple spatial scales with material, energy, and other living organism flows into IWRM will be an important theme going forward. This theme requires interdisciplinary work that involves many fields of science. Of course, many practical activities have already begun in fields such as integrated coastal management, trans-boundary aquifers, and international rivers. One important challenge for environmental scientists might be providing systematic knowledge about how can we incorporate the global dimensions of water into local watershed management systems.

21.4.2 Future Research Agenda

We have seen that global and local water are inevitably inter-connected. The form of this inter-connection will be changing. Thus, depending upon the temporal structure of inter-connectedness, boundaries should also be changed. An important question is whether, as society inevitably adapts to newly emerging something-sheds, such as heat-sheds, nutrient-sheds, and bio-sheds, our social systems can rapidly respond and transform themselves to manage these newly emerged something-sheds. This requires that the boundaries of commons should be flexibly changed. Though Oström (2009) pointed out that boundaries are important for the sustainable management of commons, they did not offer any insight on the case of boundary change. Seki (2005) also pointed out that their discussion did not mention necessary conditions for the generation of commons. Thus, trans-disciplinary work related to water should concentrate on how we can generate flexible commons.

This idea is consistent with the adaptive management concept (Holling 1978; Gunderson 1999), which originally emerged from ecology. Moreover, we can find a similar concept in the field of information technology. Tokoro (2010) advocated the forthcoming scientific mission of 'open systems science'. We can find two important characteristics in the problems open system science targets. One is that there are interactions between subsystems and the larger system that includes subsystems. The other is that problems should be treated as living forms. This means that we only have one chance to solve problems. Problem awareness stems from the limitations of reductionism. Actually, analysis by reducing a problem into a simpler problems and synthesizing results has been the two important axes of scientific investigation. They pointed out that what is insufficient in conventional science is 'management'. Usually, management has been considered as outside of science. It has been thought of as a sort of personal experience heavily dependent upon personal skills (Polanyi 1974). Thus, the methodology of open systems science is defined by including 'management' as a third index in scientific activities in addition to 'analysis' and 'synthesis' When we consider 'management' of systems, especially for human made systems, Tokoro (2010) pointed out that the concepts of 'operation', 'maintenance', 'improvement', and 'change' are important.

As mentioned in Sect. 21.2, many boundaries found in the world are intimately related to social systems such as laws, policies, and economies. Thus, adjustment and integration of these fragmented entities should be considered. As for social institutions, the sectionalism commonly observed in many social institutions has a negative influence on boundary problems, increasing the inflexibility of each fragmented section. To alleviate sectionalism problems, Star and Griesemer (1989) suggested the importance of 'boundary objects'. Max-Neef (2005) also proposed a similar concept. Boundary objects are a kind of common object shared by all people who belong to different sections or research fields. Boundary object can be instruments, normalized standards, concepts, theories, and so on. An important starting point in cultivating shared boundary objects is that there is no need for only one strict definition. Instead, the diverse meanings that each person adds to the object should be accepted. Indeed, some studies of water negotiation have attempted to find boundary objects (Kistin and Ashton 2008). Thus, the fuzzy sharing of boundary objects is central to the boundary object concept. This perspective can shed new light on the dilemma of watersheds.

21.5 Summary

The latest scientific findings urge us to reconsider the conventional watershed concept. The conventional watershed concept is valuably applied in IWM, as it gives us definite boundaries based on an objective standard. However, we should be conscious of the fact that every boundary human beings draw is temporal. If we accept that every scientific activity is also a part of activities of human beings, boundaries used in scientific definition cannot escape the 'dilemma of boundaries'. This means that we can't add special privilege to boundaries scientifically defined. In some cases, conventional watershed boundaries work well, but in some cases, they may not work well at all.

Since global and local water are inevitably inter-connected and the structure of inter-connectedness is changing, boundaries should change according to the structural change of inter-connectedness. Thus, the ability of our social systems to adapt to newly emerging boundaries is important. This idea is compatible with the adaptive management concept which emerged from ecology, and similar ideas can also be found in the field of information technology. The common theme found in both fields is 'management'. Though 'management' has not been thought of as a true scientific objective, particularly in the natural sciences, these sciences should, with the aid of the social sciences, include 'management' as a challenging objective.

References

Allan JA (1999) Water stress and global mitigation: water, food and trade. Arid Land Newslett 45

Dirmeyer PA, Brubaker KL, DelSole T (2009) Import and export of atmospheric water vapor between nations. J Hydrology 365:11–22

Gunderson LH (1999) Resilience, flexibility, and adaptive management: antidotes for spurious certitude? Conserv Ecol 3(1):7

Hardin G (1968) Tragedy of commons. Science 162:1243–1248

Holling CS (1978) Adaptive environmental assessment and management. Blackbum Press, Caldwell, NJ

Kistin EJ, Ashton PJ (2008) Adapting to change in transboundary rivers: an analysis of treaty flexibility on the Orange-Senqu river basin. Int J Water Resour Dev 24(3):385–400

Leeks JGL, Jarvie HP (1998) Introduction to the land–ocean interaction study (LOIS): rationale and international context. Sci Total Environ 210:5–20

Max-Neef MA (2005) Foundations of transdisciplinarity. Ecol Econ 53:5–16

Oki T, Kanae S (2008) Global hydrological cycles and world water resources. Science 313:1068–1072

Oström E (2009) A general framework for analyzing sustainability of social-ecological systems. Science 325:419–422. doi:10.1126/science.1172133

Polanyi M (1974) Personal knowledge—towards a post-critical philosophy. The University of Chicago Press, London, p 442

RIHN (2006) RIHN 1st international symposium proceedings—water and better human life in the future, pp 142

Rockström J, Steffen W, Noone K, Persson Å, Chapin FS, Lambin EF, Lenton TM, Scheffer M, Folke C, Schellnhuber HJ, Nykvist B, de Wit CA, Hughes T, van der Leeuw S, Rodhe H, Sörlin S, Snyder PK, Costanza R, Svedin U, Falkenmark M, Karlberg L, Corell RW, Fabry VJ, Hansen J, Walker B, Liverman D, Richardson K, Crutzen P, Foley JA (2009) A safe operating space for humanity. Nature 461:472–475

Seki Y (2005) Restoration of rainforest as a complex adaptive system—transition from unauthorized deforestation to sustainable forestry. Ochanomizu Shuppan, Tokyo (in Japanese)

Star SL, Griesemer JR (1989) Institutional ecology, translations and boundary objects: amateurs and professionals in Berkeley's museum of vertebrate zoology, 1907–39. Soc Stud Sci 19:387–420

Tokoro M (ed) (2010) Open systems science—from understanding principles to solving problems. IOS Press, Amsterdam, p 284. ISBN 978-1-60750-468-9

Turton AR, Hattingh H, Claassen M, Roux D, Ashton P (2005) Towards a model for ecosystem governance: an integrated water resource management example. In proceedings: international symposium on ecosystems governance, KwaMaritane, South Africa, 10–13 October 2005

van der Ent RJ, Savenije HHG, Schaefli B, Steele-Dunne SC (2010) Origin and fate of atmospheric moisture over continents. Water Resour Res 46:W09525. doi:10.1029/2010WR009127

Resolution Adopted by the General Assembly

[on the report of the Sixth Committee (A/63/439)]
63/124. The law of transboundary aquifers

The General Assembly,

Having considered chapter IV of the report of the International Law Commission on the work of its sixtieth session, which contains the draft articles on the law of transboundary aquifers,

Noting that the Commission decided to recommend to the General Assembly (a) to take note of the draft articles on the law of transboundary aquifers in a resolution, and to annex the articles to the resolution; (b) to recommend to States concerned to make appropriate bilateral or regional arrangements for the proper management of their transboundary aquifers on the basis of the principles enunciated in the articles; and (c) to also consider, at a later stage, and in view of the importance of the topic, the elaboration of a convention on the basis of the draft articles,

Emphasizing the continuing importance of the codification and progressive development of international law, as referred to in Article 13, paragraph 1 (a), of the Charter of the United Nations,

Noting that the subject of the law of transboundary aquifers is of major importance in the relations of States,

Taking note of the comments of Governments and the discussion in the Sixth Committee at the sixty-third session of the General Assembly on this topic,

1. *Welcomes* the conclusion of the work of the International Law Commission on the law of transboundary aquifers and its adoption of the draft articles and a detailed commentary on the subject;
2. *Expresses* its appreciation to the Commission for its continuing contribution to the codification and progressive development of international law;
3. *Also expresses* its appreciation to the International Hydrological Programme of the United Nations Educational, Scientific and Cultural Organization and to other relevant organizations for the valuable scientific and technical assistance rendered to the International Law Commission;

M. Taniguchi and T. Shiraiwa (eds.), *The Dilemma of Boundaries: Toward a New Concept of Catchment*, Global Environmental Studies, DOI 10.1007/978-4-431-54035-9, © Springer Japan 2012

4. *Takes note* of the draft articles on the law of transboundary aquifers, presented by the Commission, the text of which is annexed to the present resolution, and commends them to the attention of Governments without prejudice to the question of their future adoption or other appropriate action;
5. *Encourages* the States concerned to make appropriate bilateral or regional arrangements for the proper management of their transboundary aquifers, taking into account the provisions of these draft articles;
6. *Decides* to include in the provisional agenda of its sixty-sixth session an item entitled "The law of transboundary aquifers" with a view to examining, inter alia, the question of the form that might be given to the draft articles.

67th Plenary Meeting, 11 December 2008

Annex

The law of transboundary aquifers

Conscious of the importance for humankind of life-supporting groundwater resources in all regions of the world,

Bearing in mind Article 13, paragraph 1 (a), of the Charter of the United Nations, which provides that the General Assembly shall initiate studies and make recommendations for the purpose of encouraging the progressive development of international law and its codification,

Recalling General Assembly resolution 1803 (XVII) of 14 December 1962 on permanent sovereignty over natural resources,

Reaffirming the principles and recommendations adopted by the United Nations Conference on Environment and Development of 1992 in the Rio Declaration on Environment and Development and Agenda 21,

Taking into account increasing demands for freshwater and the need to protect groundwater resources,

Mindful of the particular problems posed by the vulnerability of aquifers to pollution,

Convinced of the need to ensure the development, utilization, conservation, management and protection of groundwater resources in the context of the promotion of the optimal and sustainable development of water resources for present and future generations,

Affirming the importance of international cooperation and good-neighbourliness in this field,

Emphasizing the need to take into account the special situation of developing countries,

Recognizing the necessity to promote international cooperation,

Part one
Introduction

Article 1

Scope

The present articles apply to:

(a) Utilization of transboundary aquifers or aquifer systems;
(b) Other activities that have or are likely to have an impact upon such aquifers or aquifer systems; and
(c) Measures for the protection, preservation and management of such aquifers or aquifer systems.

Article 2

Use of terms

For the purposes of the present articles:

(a) "aquifer" means a permeable water bearing geological formation underlain by a less permeable layer and the water contained in the saturated zone of the formation;
(b) "aquifer system" means a series of two or more aquifers that are hydraulically connected;
(c) "transboundary aquifer" or "transboundary aquifer system" means, respectively, an aquifer or aquifer system, parts of which are situated in different States;
(d) "aquifer State" means a State in whose territory any part of a transboundary aquifer or aquifer system is situated;

(e) "utilization of transboundary aquifers or aquifer systems" includes extraction of water, heat and minerals, and storage and disposal of any substance;
(f) "recharging aquifer" means an aquifer that receives a non-negligible amount of contemporary water recharge;
(g) "recharge zone" means the zone which contributes water to an aquifer, consisting of the catchment area of rainfall water and the area where such water flows to an aquifer by run-off on the ground and infiltration through soil;
(h) "discharge zone" means the zone where water originating from an aquifer flows to its outlets, such as a watercourse, a lake, an oasis, a wetland or an ocean.

Part two
General principles

Article 3

Sovereignty of aquifer States

Each aquifer State has sovereignty over the portion of a transboundary aquifer or aquifer system located within its territory. It shall exercise its sovereignty in accordance with international law and the present articles.

Article 4

Equitable and reasonable utilization

Aquifer States shall utilize transboundary aquifers or aquifer systems according to the principle of equitable and reasonable utilization, as follows:

(a) They shall utilize transboundary aquifers or aquifer systems in a manner that is consistent with the equitable and reasonable accrual of benefits therefrom to the aquifer States concerned;
(b) They shall aim at maximizing the long-term benefits derived from the use of water contained therein;
(c) They shall establish individually or jointly a comprehensive utilization plan, taking into account present and future needs of, and alternative water sources for, the aquifer States; and
(d) They shall not utilize a recharging transboundary aquifer or aquifer system at a level that would prevent continuance of its effective functioning.

Article 5

Factors relevant to equitable and reasonable utilization

1. Utilization of a transboundary aquifer or aquifer system in an equitable and reasonable manner within the meaning of article 4 requires taking into account all relevant factors, including:
   (a) The population dependent on the aquifer or aquifer system in each aquifer State;
   (b) The social, economic and other needs, present and future, of the aquifer States concerned;
   (c) The natural characteristics of the aquifer or aquifer system;
   (d) The contribution to the formation and recharge of the aquifer or aquifer system;
   (e) The existing and potential utilization of the aquifer or aquifer system;
   (f) The actual and potential effects of the utilization of the aquifer or aquifer system in one aquifer State on other aquifer States concerned;
   (g) The availability of alternatives to a particular existing and planned utilization of the aquifer or aquifer system;
   (h) The development, protection and conservation of the aquifer or aquifer system and the costs of measures to be taken to that effect;
   (i) The role of the aquifer or aquifer system in the related ecosystem.
2. The weight to be given to each factor is to be determined by its importance with regard to a specific transboundary aquifer or aquifer system in comparison with that of other relevant factors. In determining what is equitable and reasonable utilization, all relevant factors are to be considered together and a conclusion reached on the basis of all the factors. However, in weighing different kinds of utilization of a transboundary aquifer or aquifer system, special regard shall be given to vital human needs.

Article 6

Obligation not to cause significant harm

1. Aquifer States shall, in utilizing transboundary aquifers or aquifer systems in their territories, take all appropriate measures to prevent the causing of significant harm to other aquifer States or other States in whose territory a discharge zone is located.
2. Aquifer States shall, in undertaking activities other than utilization of a transboundary aquifer or aquifer system that have, or are likely to have, an impact

upon that transboundary aquifer or aquifer system, take all appropriate measures to prevent the causing of significant harm through that aquifer or aquifer system to other aquifer States or other States in whose territory a discharge zone is located.

3. Where significant harm nevertheless is caused to another aquifer State or a State in whose territory a discharge zone is located, the aquifer State whose activities cause such harm shall take, in consultation with the affected State, all appropriate response measures to eliminate or mitigate such harm, having due regard for the provisions of articles 4 and 5.

Article 7

General obligation to cooperate

1. Aquifer States shall cooperate on the basis of sovereign equality, territorial integrity, sustainable development, mutual benefit and good faith in order to attain equitable and reasonable utilization and appropriate protection of their transboundary aquifers or aquifer systems.
2. For the purpose of paragraph 1, aquifer States should establish joint mechanisms of cooperation.

Article 8

Regular exchange of data and information

1. Pursuant to article 7, aquifer States shall, on a regular basis, exchange readily available data and information on the condition of their transboundary aquifers or aquifer systems, in particular of a geological, hydrogeological, hydrological, meteorological and ecological nature and related to the hydrochemistry of the aquifers or aquifer systems, as well as related forecasts.
2. Where knowledge about the nature and extent of a transboundary aquifer or aquifer system is inadequate, aquifer States concerned shall employ their best efforts to collect and generate more complete data and information relating to such aquifer or aquifer system, taking into account current practices and standards. They shall take such action individually or jointly and, where appropriate, together with or through international organizations.
3. If an aquifer State is requested by another aquifer State to provide data and information relating to an aquifer or aquifer system that are not readily available, it shall employ its best efforts to comply with the request. The requested State may

condition its compliance upon payment by the requesting State of the reasonable costs of collecting and, where appropriate, processing such data or information.
4. Aquifer States shall, where appropriate, employ their best efforts to collect and process data and information in a manner that facilitates their utilization by the other aquifer States to which such data and information are communicated.

Article 9

Bilateral and regional agreements and arrangements

For the purpose of managing a particular transboundary aquifer or aquifer system, aquifer States are encouraged to enter into bilateral or regional agreements or arrangements among themselves. Such agreements or arrangements may be entered into with respect to an entire aquifer or aquifer system or any part thereof or a particular project, programme or utilization except insofar as an agreement or arrangement adversely affects, to a significant extent, the utilization by one or more other aquifer States of the water in that aquifer or aquifer system, without their express consent.

Part three
Protection, preservation and management

Article 10

Protection and preservation of ecosystems

Aquifer States shall take all appropriate measures to protect and preserve ecosystems within, or dependent upon, their transboundary aquifers or aquifer systems, including measures to ensure that the quality and quantity of water retained in an aquifer or aquifer system, as well as that released through its discharge zones, are sufficient to protect and preserve such ecosystems.

Article 11

Recharge and discharge zones

1. Aquifer States shall identify the recharge and discharge zones of transboundary aquifers or aquifer systems that exist within their territory. They shall take appropriate measures to prevent and minimize detrimental impacts on the recharge and discharge processes.
2. All States in whose territory a recharge or discharge zone is located, in whole or in part, and which are not aquifer States with regard to that aquifer or aquifer system, shall cooperate with the aquifer States to protect the aquifer or aquifer system and related ecosystems.

Article 12

Prevention, reduction and control of pollution

Aquifer States shall, individually and, where appropriate, jointly, prevent, reduce and control pollution of their transboundary aquifers or aquifer systems, including through the recharge process, that may cause significant harm to other aquifer States. Aquifer States shall take a precautionary approach in view of uncertainty about the nature and extent of a transboundary aquifer or aquifer system and of its vulnerability to pollution.

Article 13

Monitoring

1. Aquifer States shall monitor their transboundary aquifers or aquifer systems. They shall, wherever possible, carry out these monitoring activities jointly with other aquifer States concerned and, where appropriate, in collaboration with competent international organizations. Where monitoring activities cannot be carried out jointly, the aquifer States shall exchange the monitored data among themselves.
2. Aquifer States shall use agreed or harmonized standards and methodology for monitoring their transboundary aquifers or aquifer systems. They should identify key parameters that they will monitor based on an agreed conceptual model of the aquifers or aquifer systems. These parameters should include parameters on the condition of the aquifer or aquifer system as listed in article 8, paragraph 1, and also on the utilization of the aquifers or aquifer systems.

Article 14

Management

Aquifer States shall establish and implement plans for the proper management of their transboundary aquifers or aquifer systems. They shall, at the request of any of them, enter into consultations concerning the management of a transboundary aquifer or aquifer system. A joint management mechanism shall be established, wherever appropriate.

Article 15

Planned activities

1. When a State has reasonable grounds for believing that a particular planned activity in its territory may affect a transboundary aquifer or aquifer system and thereby may have a significant adverse effect upon another State, it shall, as far as practicable, assess the possible effects of such activity.
2. Before a State implements or permits the implementation of planned activities which may affect a transboundary aquifer or aquifer system and thereby may have a significant adverse effect upon another State, it shall provide that State with timely notification thereof. Such notification shall be accompanied by available technical data and information, including any environmental impact assessment, in order to enable the notified State to evaluate the possible effects of the planned activities.
3. If the notifying and the notified States disagree on the possible effect of the planned activities, they shall enter into consultations and, if necessary, negotiations with a view to arriving at an equitable resolution of the situation. They may utilize an independent fact-finding body to make an impartial assessment of the effect of the planned activities.

Part four
Miscellaneous provisions

Article 16

Technical cooperation with developing States

States shall, directly or through competent international organizations, promote scientific, educational, technical, legal and other cooperation with developing States for the protection and management of transboundary aquifers or aquifer systems, including, inter alia:

(a) Strengthening their capacity-building in scientific, technical and legal fields;
(b) Facilitating their participation in relevant international programmes;
(c) Supplying them with necessary equipment and facilities;
(d) Enhancing their capacity to manufacture such equipment;
(e) Providing advice on and developing facilities for research, monitoring, educational and other programmes;
(f) Providing advice on and developing facilities for minimizing the detrimental effects of major activities affecting their transboundary aquifer or aquifer system;
(g) Providing advice in the preparation of environmental impact assessments;
(h) Supporting the exchange of technical knowledge and experience among developing States with a view to strengthening cooperation among them in managing the transboundary aquifer or aquifer system.

Article 17

Emergency situations

1. For the purpose of the present article, "emergency" means a situation, resulting suddenly from natural causes or from human conduct, that affects a transboundary

aquifer or aquifer system and poses an imminent threat of causing serious harm to aquifer States or other States.

2. The State within whose territory the emergency originates shall:
   (a) Without delay and by the most expeditious means available, notify other potentially affected States and competent international organizations of the emergency;
   (b) In cooperation with potentially affected States and, where appropriate, competent international organizations, immediately take all practicable measures necessitated by the circumstances to prevent, mitigate and eliminate any harmful effect of the emergency.
3. Where an emergency poses a threat to vital human needs, aquifer States, notwithstanding articles 4 and 6, may take measures that are strictly necessary to meet such needs.
4. States shall provide scientific, technical, logistical and other cooperation to other States experiencing an emergency. Cooperation may include coordination of international emergency actions and communications, making available emergency response personnel, emergency response equipment and supplies, scientific and technical expertise and humanitarian assistance.

Article 18

Protection in time of armed conflict

Transboundary aquifers or aquifer systems and related installations, facilities and other works shall enjoy the protection accorded by the principles and rules of international law applicable in international and non-international armed conflict and shall not be used in violation of those principles and rules.

Article 19

Data and information vital to national defence or security

Nothing in the present articles obliges a State to provide data or information vital to its national defence or security. Nevertheless, that State shall cooperate in good faith with other States with a view to providing as much information as possible under the circumstances.

Index

M. Taniguchi and T. Shiraiwa (eds.), *The Dilemma of Boundaries: Toward a New Concept of Catchment*, Global Environmental Studies, DOI 10.1007/978-4-431-54035-9, © Springer Japan 2012

MIX
Papier aus verantwortungsvollen Quellen
Paper from responsible sources
FSC® C105338
FSC
www.fsc.org

If you have any concerns about our products,
you can contact us on
ProductSafety@springernature.com

In case Publisher is established outside the EU,
the EU authorized representative is:
Springer Nature Customer Service Center GmbH
Europaplatz 3, 69115 Heidelberg, Germany

Printed by Libri Plureos GmbH
in Hamburg, Germany